Soil Organic Carbon Sequestration in Terrestrial Biomes of the United States

Klaus Lorenz • Rattan Lal

Soil Organic Carbon Sequestration in Terrestrial Biomes of the United States

Springer

Klaus Lorenz
Assistant Director
CFAES Rattan Lal Center for Carbon
Management and Sequestration
Columbus, OH, USA

Rattan Lal
Director
CFAES Rattan Lal Center for Carbon
Management and Sequestration
Columbus, OH, USA

ISBN 978-3-030-95192-4 ISBN 978-3-030-95193-1 (eBook)
https://doi.org/10.1007/978-3-030-95193-1

This Springer imprint is published by the registered company Springer Nature Switzerland AG
The registered company address is: Gewerbestrasse 11, 6330 Cham, Switzerland

Preface

The terrestrial biomes of the USA contain soil organic carbon (SOC) stocks determined by the balance between carbon (C) input and output, resulting in site-specific equilibrium levels maintained under natural conditions in the long term. Soil quality and health and many terrestrial ecosystem services related to climate change adaptation, food security, and biodiversity due to increased soil fertility, improved moisture retention, improved water availability to plants, reduced soil erosion, and improved habitat for biodiversity are supported by an adequate, site-specific SOC stock. However, terrestrial biomes of the USA have lost SOC that had been accumulated over centuries mainly by conversion of native vegetation (forests and grasslands) to cropland agriculture, besides other processes, resulting in soil degradation. Thus, there is the potential to draw down some of the C lost as carbon dioxide (CO_2) from the atmosphere by soil and land-use management practices through implementing climate-resilient crop-, forest-, grass-, range-, and wetland practices. There is now a wealth of knowledge available on SOC-positive soil and land-use management practices. The increases in SOC stocks in response to implementing those practices will not only improve environmental quality but also contribute to climate change adaptation and mitigation as SOC is the largest active terrestrial pool of organic C. However, adjustments in management practices to maintain and/or increase SOC stocks come at a cost to farmers, forest landowners, and ranchers. The emerging voluntary carbon market may be a possibility to reward the implementation of climate-smart practices by generating carbon credits and carbon offsets as many corporations in the USA announced goals to reduce their carbon footprint, achieve climate neutrality, and net-zero emissions. Regulatory carbon markets on state and federal level, on the other hand, are in a nascent state, and their future is uncertain as they depend on continued support which is subject to political decisions. Among major issues in any carbon market that need to be addressed are: (i) accuracy, affordability, and availability of data for measurement, reporting, verification, and monitoring of changes in SOC stocks, (ii) inferring such SOC stock changes based on the implementation of management practices, and (iii) use of data published from some long-term experiments established in the USA since 1960s. This book has collated and synthesized the available information. There is also the need for information on economic benefits for farmers, forest landowners, and ranchers to support the implementation of practices related to SOC stock increases. Alternative to carbon markets, providing consumers with information to choose climate-smart products indicated by a seal similar to the successful United States Department of Agriculture seal for organic products may be an opportunity to reward land managers for SOC-sequestering practices based on price premiums on those products.

This book presents terrestrial biomes of the USA, their SOC stocks, how natural disturbances affect them, and how management interventions can increase SOC sequestration. The often-overlooked importance of the soil inorganic carbon (SIC) stock and its dynamics is also discussed. The book concludes with a chapter on how climate-smart and particularly SOC-smart soil and land-use management practices may be incentivized. We want to thank all researchers in the USA and around the globe who contribute to improving our understanding of the processes resulting in the build-up and maintenance of the soil C stock as the basis for identifying soil C sequestering practices.

This project was supported through funding by the CFAES Rattan Lal Center for Carbon Management and Sequestration (C-MASC) at the Office of Research of the Ohio State University.

Columbus, OH, USA Klaus Lorenz

Columbus, OH, USA Rattan Lal
August 31, 2021

Contents

Chapter 1
Terrestrial Land of the United States of America

Abstract The terrestrial land area of the United States of America (U.S.) excluding overseas territories covers about 9.6 million km^2 (3.7 million mi^2). In 2012, grassland, pasture and rangeland covered an estimated 2.7 million km^2 (1.0 million mi^2), forest land 2.6 million km^2 (1.0 million mi^2), and cropland 1.6 million km^2 (0.6 million mi^2). The total developed area including urban areas in the conterminous U.S. (CONUS), i.e., the 48 adjoining U.S. states on the continent of North America, was estimated at 430,000 km^2 (165,000 mi^2) in the year 2016. Aside land use, the terrestrial biosphere of the U.S. can also be divided into biomes which can be defined as systems containing multiple ecosystems that have common physical properties such as climate and geology which all affect soil organic carbon (SOC) sequestration and stocks. Forest biomes are important carbon (C) sinks with more than half of the forest C stock found in soils. Among forest biomes in the U.S. is the boreal forest/taiga biome which covers much of Alaska. This biome is characterized by acidic, cool and moist conditions, sometimes underlain by permafrost. Adverse environmental conditions together with the predominance of conifer trees contribute to the accumulation of boreal forest SOC. A comprehensive inventory of the boreal forest area is lacking. In comparison, the temperate coniferous forest biome in the U.S. is found on more fertile soils under more favorable climate, such as coniferous forests in the Pacific Northwest. Plant biomass accounts for a large portion of the total C stock but SOC stocks can also be dominant. An area of about 1.3 million km^2 (0.5 million mi^2) in the U.S. is under temperate coniferous forest. Similarly favored by environmental conditions is the temperate broadleaf and mixed forest biome, covering about 1.2 million km^2 (0.5 million mi^2). Hardwoods of widely varying species dominate most of the eastern U.S. temperate broadleaf and mixed forests while conifers are more prominent in the southeastern Coastal Plain pinelands, in parts of the southeast, west and, especially, the Pacific Northwest. The tropical forest biome covers a small area of about 13,000 km^2 (5000 mi^2), mainly in Hawaii and Puerto Rico. Soils are often infertile but the tropical climate favors tree growth and decomposition and, thus, most C is stored in trees. The temperate grassland, savanna, and shrubland biome covers about 2.6 million km^2 (1.0 million mi^2) in CONUS, most of it in the Great Plains. This biome includes pastures and rangelands. Grassland plants allocate a large proportion of C to roots to adapt to dry cli-

K. Lorenz, R. Lal, *Soil Organic Carbon Sequestration in Terrestrial Biomes of the United States*, https://doi.org/10.1007/978-3-030-95193-1_1

mates, and, thus, SOC stocks are relatively large. The coldest terrestrial biome in the U.S., i.e., the tundra biome is located at high latitudes and altitudes, sometimes underlain by permafrost. Plant growth is limited by very low temperatures, short growing seasons, low nutrient availability and waterlogging. The extent of the arctic and alpine tundra biomes is uncertain. Terrestrial wetlands occur in boreal, temperate, and tropical climate zones in the U.S., and include peatlands. Here, SOC accumulates under waterlogged conditions. About 0.40 million km^2 (0.15 million mi^2) of U.S. land is covered by terrestrial wetlands, and 0.25 million km^2 (0.10 million mi^2) may be under peatland. Plant growth in deserts and xeric shrublands is limited by low annual precipitation, and phreatophytes, i.e., plants whose deep root systems penetrate into permanent sources of water thrive. This biome is located in colder (e.g., Great Basin shrub steppe) and in warmer regions (e.g., Mojave desert). Calcium carbonate deep within the soil profile may contribute to the total soil C stock. The land covers "Barren" and "Shrub/Scrub" cover about 54,000 km^2 (21,000 mi^2) and 1.3 million km^2 (0.5 million mi^2), respectively. Estimates for the U.S. desert area are uncertain as deserts are sometimes considered rangelands. The major crops planted in U.S. croplands are corn (*Zea mays* L.), wheat (*Triticum* spp.), soybean (*Glycine max* [L.] Merr.), and cotton (*Gossypium* spp.). Cropland in CONUS, Alaska and Hawaii covered an estimated 1.6 million km^2 (0.6 million mi^2) in 2012. Uncertain is the extent of settlement and urban areas in the U.S. as there is no single definition for 'urban'. The total developed area in CONUS increased by 6.7% between 2001 and 2016. Aside for urban uses, human activities restructure the terrestrial biosphere of the U.S. for agriculture, forestry, and other uses. This transformation substantially alters the biogeochemical cycles and affects SOC stocks aside natural disturbances, climate and global changes. This chapter presents terrestrial biomes of the U.S., and discuss the effects of human activities.

Keywords Ecosystem · Terrestrial biomes · Conterminous United States · Boreal forest biome · Temperate coniferous forest biome · Temperate broadleaf and mixed forest biome · Tropical forest biome · Temperate grasslands, savannas, and shrublands biome · Tundra biome · Terrestrial wetlands · Desert and xeric shrubland biome · Croplands · Dense settlements and villages

1.1 Introduction

An ecosystem can be defined as the interacting community of organisms and non-living components such as minerals, nutrients, water, weather and topographic features in a specific environment (Cavicchioli et al. 2019). Ecosystems combine the abiotic environment with biological communities (plants, animals, fungi, microorganisms) that form self-regenerative functional units (Barrett et al. 2020). Systems containing multiple ecosystems that have common physical properties such as climate and geology can be defined as biomes. This is based on Schimper's concept that plant growth forms are distributed across the globe in accordance with the

Table 1.1 Land area, soil organic and inorganic carbon stocks, and carbon densities of terrestrial biomes/land cover types of the conterminous United States (soil organic carbon stocks based on Liu et al. 2020b, soil inorganic carbon stocks based on Guo et al. 2006; estimates for soil organic and inorganic carbons stocks are not directly comparable as different approaches were used)

Biome/Land cover type	Area		Soil organic carbon stock		Soil inorganic carbon stock[a]		Total carbon density	
	million km^2	million mi^2	Pg C	Billion tn C	Pg C	Billion tn C	kg C m^{-2}	lbs C ft^{-2}
Forest	2.7	1.0	40.5	44.6	4.6	5.0	22.8	4.7
Cropland	2.0	0.8	28.4	31.3	15.1	16.6	15.9	3.3
Shrubland	1.7	0.6	11.0	12.1	15.4	17.0	8.0	1.6
Grassland	1.2	0.5	14.7	16.2	17.5	19.3	13.6	2.8

[a]To 200 cm (79-in) depth; Cropland: Terrestrial ecosystem 'Agriculture', Shrubland: Terrestrial ecosystem 'Shrub', Grassland: Sum of terrestrial ecosystems 'Grass' and 'Pasture' based on Guo et al. (2006)

availability of light, water, nutrients and heat (Schimper 1898). Terrestrial biomes delineate major types of ecological associations that occupy broad geographic regions of land (Campbell and Reece 2005). They arise as combinations of dominant plant types with seventeen terrestrial biomes differentiated by Prentice et al. (1992). Terrestrial biomes describe patterns of ecosystem form, process and biodiversity. Delimiting biomes is, however, not objective as it relies on expert knowledge (Conradi et al. 2020). Ecological and evolutionary questions that a study aims to answer will define what kind of biome concept is applied. Thus, biome maps are constructed in different ways (Mucina 2019). Biomes are basic building blocks that make up the biosphere. Biome is a multiscale phenomenon, spanning several large-scale spatial levels, including global climatic zones, continents and landscapes at subcontinental and supraregional scales. Biomes may show various vegetation physiognomic aspects that could represent multiple stable states. The patches of biomes are linked by a common network of ecological processes that define the selective pressures as macroclimatic, soil-related, hydrological and natural large-scale disturbance factors and stressors (Mucina 2019).

This book collates and summarizes data on soil organic carbon (SOC) stocks for terrestrial biomes of the United States of America (U.S.), and discusses options for enhancing SOC sequestration in managed areas of the terrestrial biomes. One chapter is devoted to soil inorganic carbon (SIC) stocks for terrestrial biomes of the U.S. (Table 1.1).

1.2 Terrestrial Land Area

The potential for SOC sequestration in terrestrial ecosystems of the U.S. can be assessed in detail by differentiating among land uses and covers. The 'total' area of the U.S. (50 states and the District of Columbia, excluding overseas territories) is

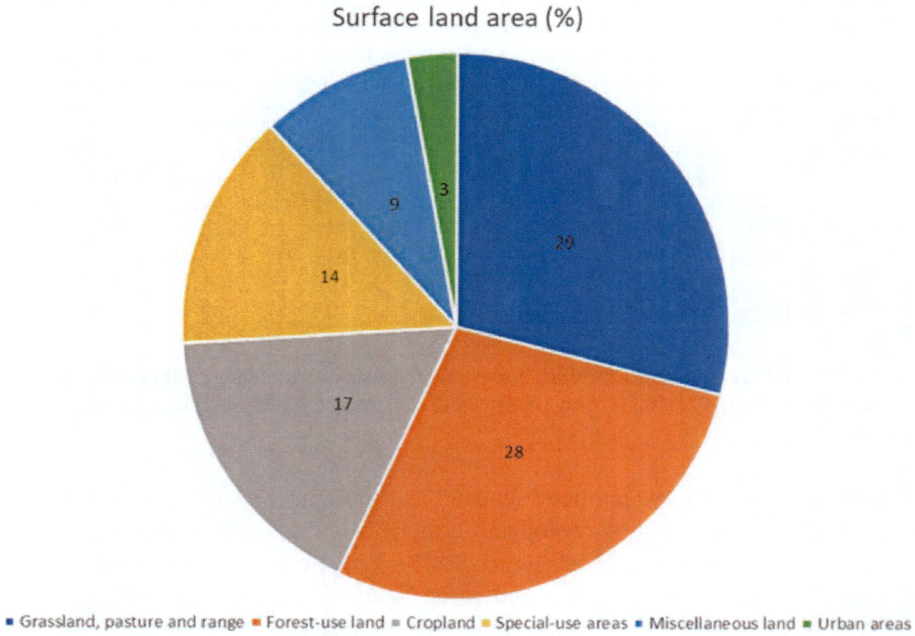

Fig. 1.1 Surface land area by land use in percent of total land area for the conterminous 48 U.S. states (CONUS, i.e., the 48 adjoining U.S. states on the continent of North America), Alaska and Hawaii in the year 2012. (Bigelow and Borchers 2017)

estimated at about 9.3–9.8 million km^2 (3.6–3.8 million mi^2; Bigelow and Borchers 2017; Central Intelligence Agency 2020). The populated territories of Puerto Rico, American Samoa, Guam, Northern Mariana Islands, and U.S. Virgin Islands together cover another 23,789 km^2 (9185 mi^2; U.S. Census Bureau 2012).

Two primary sources of data used to define land use and land cover of the conterminous U.S. (CONUS), i.e., the 48 adjoining U.S. states on the continent of North America, are the National Resources Inventory (NRI) and the National Land Cover Database (NLCD; Nelson et al. 2020). Each provides different perspectives of changes in the U.S. land base. The NRI collects data and produces an array of information related to land use, land cover, soil conditions, conservation practices, and other attributes on non-Federal lands. The NLCD comprises consistent land cover information at the national scale for a wide variety of environmental, land management, and modeling applications (Nelson et al. 2020). In 2012, the largest land areas of CONUS plus Alaska and Hawaii in decreasing order were grassland pasture and range (2.7 million km^2 [1.0 million mi^2]), forest-use land (2.6 million km^2 [1.0 million mi^2]), and cropland (1.6 million km^2 [0.6 million mi^2]; Fig. 1.1; Bigelow and Borchers 2017). A total of 41% of land area of CONUS alone is used to feed livestock, i.e., 10% for growing feed crops and the remainder for grazing cattle (*Bos bos taurus* Linnaeus, 1758; Merrill and Leatherby 2018).

The cropland area includes five components: cropland harvested, crop failure, cultivated summer fallow, cropland used only for pasture, and idle cropland

(Bigelow and Borchers 2017). Cropland harvested includes row crops and closely sown crops; hay and silage crops; tree fruits, small fruits, berries, and tree nuts; vegetables and melons; and miscellaneous other minor crops. Crop failure consists mainly of the acreage on which crops failed because of weather, insects, and diseases but does include some land not harvested. Cultivated summer fallow refers to cropland in sub-humid regions of the West that are cultivated for one or more seasons to control weeds and store moisture before small grains are planted. Cropland pasture is considered to be in long-term crop rotation. Idle cropland includes land in cover and soil-improvement crops, and cropland on which no crops were planted. The land-use grassland pasture and range encompasses all open land used primarily for pasture and grazing, including shrub and brush land types of pasture; grazing land with sagebrush (*Artemisia tridentate*) and scattered mesquite (*Prosopis* spp.); and all tame and native grasses, legumes, and other forage used for pasture or grazing (Bigelow and Borchers 2017).

The forest-use land category includes both grazed and un-grazed forests but excludes forestland in parks, wildlife areas, and similar special-purpose uses (Bigelow and Borchers 2017). Special-use areas include highways, roads, and railroad rights-of-way and airports; Federal and State parks, wilderness areas, and wildlife refuges; national defense and industrial areas; and farmsteads and farm roads. Urban areas include densely populated areas with at least 50,000 people ("urbanized areas") and densely populated areas with 2500–50,000 people ("urban clusters"). Miscellaneous other land includes other uses, such as industrial and commercial sites in rural areas, cemeteries, golf courses, mining areas, quarry sites, marshes, swamps, sand dunes, bare rocks, deserts, tundra, rural residential, and other unclassified land (Bigelow and Borchers 2017).

Freedgood et al. (2020) created baseline land cover/use maps for the agricultural landscape of CONUS for the year 2016. Private agricultural land covered 3.7 million km^2 (1.4 million mi^2) of cropland, pastureland, rangeland, and woodland associated with farms. Another 0.9 million km^2 (0.3 million mi^2) of federal land was leased for grazing. The agricultural land of CONUS is part of a larger mosaic of land covers and uses, including forestland, urban areas, low-density residential areas, ungrazed federal land, and unclassified rural land (Freedgood et al. 2020).

Land use and land cover in the U.S. are changing over time. Land use projections are parameterized using the NRI data because it offers the longest time trend for the non-Federal CONUS and provides information for both land use and cover (Nelson et al. 2020). However, NRI does not inventory Federal lands. The NLCD provides wall-to-wall maps of the land base and is, therefore, the data source for landscape pattern analyses, including landscape mosaic and fragmentation patterns (Nelson et al. 2020). The temporal changes can be assessed based on the NLCD product suite which includes Landsat-based, 30 m (98 ft) resolution products over CONUS for land cover, urban imperviousness, and tree, shrub, herbaceous and bare ground fractional percentages (Homer et al. 2020). Overall, land cover has declined in forests, cropland, and pasture but increased in barren, scrub/shrub, and developed classes, which are particularly concentrated in the U.S. Southeast (Wentland et al. 2020). Between 2001 and 2016, 50% of the overall land cover change in CONUS involved forest, driven by change agents of harvest, fire, disease and pests that resulted in an overall forest decline, including increasing fragmentation and loss of interior forest (Homer et al. 2020). Agricultural change represented 15.9% of the change, with a substantial decline (7.94%) in pasture/hay transitioning mostly to cultivated crop. Grass and shrub change comprised 14.5% of the total change, with most change resulting from fire. Change in developed land cover was the most persistent, and permanent land change adding almost 29,000 km^2 (11,197 mi^2) over 15 years or 5.6% of the total change in land cover of CONUS. Southern states exhibited developed land cover expansion much faster than most of the northern states (Homer et al. 2020). Over the past 150 years, about half of the wetland area in CONUS has been converted mainly by draining and conversion to agriculture (Dahl 1990).

Agricultural industrialization in the U.S. is accompanied by changes in overall cropland area as well as declines in crop diversity (Crossley et al. 2021). The total area of cropland peaked at 28% of the CONUS area in 1920, with the percent of land area in 18 major crops dropping since then to 17% by 2017. For example, the area under corn (*Zea mays* L.), wheat (*Triticum* spp.) and hay declined by 79,500, 88,500 and 110,500 km^2 (30,695, 34,170 and 42,670 mi^2) in 1940–2017, respectively. In contrast, the CONUS area under soybean (*Glycine max* [L.] Merr.), increased by 226,700 km^2 (87,529 mi^2) in the same period. In some areas, especially the Northeast, the decline in cropland area resulted in an increase in forest area while in other areas, cropland was lost due to urban sprawl (Crossley et al. 2021).

1.3 Principal Terrestrial Biomes of the U.S.

1.3.1 Forest Biomes

In the Resources Planning Act (RPA) Assessment, Forest Inventory and Analysis (FIA) data are used to determine trends in forest land across all ownerships in the U.S., providing information based on a land use perspective of forest land (Nelson et al. 2020). The forest and woodland area of the U.S occupies about 3.3 million km^2 (1.3 million mi^2) with the forest area alone covering 2.9–3.1 million km^2 (1.1–1.2 million mi^2; Oswalt et al. 2019; U.S. Environmental Protection Agency 2021). For CONUS alone, Magerl et al. (2019) estimated a forest area of 2.5 million km^2 (1.0 million mi^2). Forests largely free from logging and other forms of extraction and development (i.e., intact forests) cover 7% of CONUS forest area (Oswalt et al. 2019). In 2015, protected and difficult to access forests ("primary forests") covered 0.75 million km^2 (0.29 million mi^2) in the U.S. equivalent to 24% of the national forest area (Bernier et al. 2017). The Food and Agriculture Organization of the United Nations (FAO) defines "primary forest" as "naturally regenerated forest of native species where there are no clearly visible indications of human activities and the ecological processes are not significantly disturbed" (FAO 2006). Across the

U.S., 58% of forest and woodland area is privately owned and 42% is publicly owned (Oswalt et al. 2019). Most of the federally owned forested land lies in the 11 western contiguous states while most of the forested land in the eastern U.S. is privately owned (Liu et al. 2021).

Between 1973 and 2000, forest cover in CONUS decreased by 97,000 km^2 (37,452 mi^2; Sleeter et al. 2013). Eastern Temperate Forests lost 61,600 km^2 (23,784 mi^2), and the Western Cordillera had the second highest amount of forest loss at 25,200 km^2 (9730 mi^2). From 1982 to 2016, however, tree canopy increased in the U.S. (+15%), mostly in the eastern parts of the country (Song et al. 2018). Unlike declining forest cover in the western U.S., southeastern forests are recovering from historical disturbances or are under intensive forestry management (Birdsey et al. 2006).

In 2001–2016, total forest extent in the CONUS declined by another 63,538 km^2 (24,532 mi^2; Homer et al. 2020). Specifically, the total extents of deciduous forest, evergreen forest and mixed forest declined by about 23,716 km^2 (9157 mi^2), 39,599 km^2 (15,289 mi^2) and 223 km^2 (86 mi^2). Results also indicated a net increase of forest fragmentation from 2001 until 2016 (Homer et al. 2020).

Forests provide the most stable and the highest quality water supplies among all land uses in CONUS comprising 36% of the total land area but contributing 50% of the total surface water yield (Liu et al. 2021). Another important ecosystem service by forests is their C sink (Hayes et al. 2012). From 1971 to 2015, the average forest C sink or net biome productivity (NBP) of forests in CONUS was 129 Tg C y^{-1} (142 million tn C y^{-1}; Liu et al. 2020b). From the beginning of the 1990s until around 2000, the net annual C sink in forests of CONUS was estimated at 180 to 230 Tg C y^{-1} (198 to 254 million tn C y^{-1}; Magerl et al. 2019). The total C stock of CONUS forests including soils, litter, downed deadwood, standing deadwood, and living biomass increased by 45% from 1907 to 2012 as result of vegetation thickening rather than area expansion. In 2012, total forest C stocks were 39 Pg C (42 billion tn C) corresponding to a density of 150 Mg C ha^{-1} (67 tn C ac^{-1}) and a forest biomass C density of 71 Mg C ha^{-1} (32 tn C ac^{-1}; Magerl et al. 2019). Importantly, more than 55% of forest ecosystem C is stored in the soil (U.S. EPA 2021).

Boreal Forest/Taiga Biome

The U.S. boreal forest/taiga biome (abbreviated in the following as "boreal forest") covers much of Alaska, and parts of the northern CONUS and New England, and is characterized by a climate with strong seasonal variation (Chapman and Bolen 2015; Thiffault 2019). The boreal forest biome affects high latitude albedo and, thus, global mean temperature (Bonan 2008). Typical soils are shallow and Spodosols which develop in the cool and moist climate through the process of podzolization (Chapman and Bolen 2015). Surface horizons are depleted in clay and other fine particles but enriched in sand content. The activity of the decomposer community is inhibited because of acidic, cool and moist conditions contributing to the accumulation of OM together with the decay resistant litter. Charcoal is commonly found in the boreal forest soil profile from frequent fires (Chapman and Bolen 2015). In the northern part, Gelisols are another important soil order where

Fig. 1.2 White spruce (*Picea glauca*) taiga, Denali Highway, Alaska. (Wikimedia Commons; public domain)

subsoil can remain frozen year after year with about 40–55% of all permafrost soils found in boreal forests (Allison and Treseder 2011). Conditions in permafrost slow decomposition and, thus, permafrost zone SOC stocks have accumulated over hundreds to thousands of years. Permafrost can be defined as subsurface Earth materials (e.g., rock, soil, and ice) remaining <0 °C (32 °F) for at least two consecutive years (Schuur et al. 2018).

Low minimum temperatures and low nutrient (i.e., N) availability are major stressors for plant growth (Du et al. 2020; Prentice et al. 1992). Ubiquitous symbiotic associations with ectomycorrhizal fungi have evolved to promote boreal forest tree growth through efficient uptake of nutrients, in particular through mobilization of N via decomposition of organic substrates (Finlay et al. 2020). In addition to climate, the occurrence of permafrost and the frequency of forest fires affect vegetation patterns (Thiffault 2019). Most boreal forest plant biomass C is located aboveground. The forest canopy is dominated by relatively few species of spruce (*Picea* spp.), firs (*Abies* spp.), and other conifers. The understory species richness is also low (Thiffault 2019). Mature boreal forest ecosystems consist mainly of black spruce (*Picea marianna* Mill BSP), white spruce (*Picea glauca*; Fig. 1.2), Jack pine (*Pinus banksiana* Lamb.) and lodgepole pine (*Pinus contorta*; DeLuca and Boisvenue 2012). Other major species include tamarack (*Larix laricina*), trembling aspen (*Populus tremuloides*), paper birch (*Betula papyrifera*) and balsam poplar (*Populus balsamifera*; Thiffault 2019). After fire events, pioneer species establish including bryophytes (e.g., *Polytricum*), grasses (*Deschampsia flexuosa*), herbs (e.g., *Ebolobium*) and trees (*Betula* spp., *Populus* spp., *Picea* spp.).

Forests of Alaska were mapped as consisting of 43% softwoods, 32% aspen-birch, 17% fir-spruce, 7% hemlock-Sitka spruce (*Tsuga* spp. – *Picea sitchensis* [Bong.] Carr.), and <1% hardwoods (Johnson and Kern 2003). Other species in boreal forests in the U.S. include black crowberry (*Empetrum nigrum*) and other ericas (*Vaccinium*), the Arctic willow (*Salix arctica*), white mountain avens (*Dryas octopetala*), green alder (*Alnus viridis sinuata*) and balsam poplars (https://en.wikipedia.org/wiki/Taiga). Dwarf birches (*Betula glandulosa* and *Betula nana*) may also be present along with shrubs and sedges such as *Eriophorum angustifolium* and *Carex*, and herbs such as *Menyanthes trifoliata*, *Petasites frigidus* and *Comarum palustre*. Well-drained areas have different plant communities including black cottonwood (*Populus trichocarpa*; https://en.wikipedia.org/wiki/Taiga).

A comprehensive inventory of the boreal forest area is lacking (Birdsey et al. 2018). Zhu and McGuire (2016) estimated an area of 0.64 million km^2 (0.25 million mi^2) for regions in Alaska dominated by boreal forest land-cover types. Further, Oswalt et al. (2019) reported a forest land area for Alaska of 0.52 million km^2 (0.20 million mi^2) including 0.45 million km^2 (0.17 million mi^2) of boreal forest of interior Alaska. The total forest area in Alaska including the boreal forest area has increased from 1982 to 2015 (Liu et al. 2020a). However, large-scale inventory of boreal forests of interior Alaska has started only in the year 2016, and no data on SOC stocks for the boreal forest biome of the U.S. have been published (Pattison et al. 2018).

Temperate Coniferous Forest Biome

Temperate coniferous forests in the U.S. occur in parts of the southeast, west and, especially, the Pacific Northwest (Fig. 1.3), of the coastal U.S. (Gilliam 2016). The temperate forest biome tends to dominate under more extreme conditions (i.e., cold climates and/or infertile soils), in fire-dominated regions (e.g., southeastern Coastal Plain, western regions) and in regions with unusually long photosynthetic seasons arising from mild winters with abundant rainfall (e.g., Pacific Northwest on relatively rich soils; Chapman and Bolen 2015). Coniferous forests under Mediterranean climate where the wet season is concurrent with the cold season, and droughts for several months occur in the warm, dry season are found in California (Safford and Vallejo 2019). Common soils of temperate coniferous forests are less fertile relative to soils found beneath deciduous temperate forests, and generally include the soil orders Alfisols, Inceptisols, Entisols and Spodosols (Adams et al. 2019; Safford and Vallejo 2019). Otherwise, soils in the Pacific temperate rainforest region are generally deep, hold abundant moisture, are not subject to deep-freezing, and are enriched in extractable Fe and Al (Carpenter et al. 2014). SOC and total soil N contents are high overall but also variable. Compared to tropical rainforest soils in the Pacific Basin, Pacific temperate rainforest have greater weatherable minerals, cation-exchange capacities, SOC, and total exchangeable base cations. Further, soils of the Pacific temperate rainforest tend to be deep, acidic Al-saturated, and can have large N reservoirs (Carpenter et al. 2014). In contrast, in some areas accelerated podzolization can result in limited supplies of P and may limit forest growth (Kranabetter et al. 2020).

Fig. 1.3 Diamond Peak, Willamette National Forest, Oregon. (Wikimedia Commons; public domain)

Plant biomass accounts for a large portion of the total C stock of temperate coniferous forests, and is stored in large woody aboveground organs and in deep, coarse root systems (De Deyn et al. 2008). Temperate coniferous forests include both planted and natural stands (Adams et al. 2019). Important genera in montane conifer forests of western U.S. include fir, juniper (*Juniperus* spp.), spruce, pine (*Pinus* spp.), hemlock and Douglas-fir (*Pseudotsuga menziesii*; Gilliam 2016). In addition, several genera of cedars (family Cupressaceae, e.g., *Calocedrus*, *Chamaecyparis*, *Thuja*), including the redwood (*Sequoia sempervirens*) occur in the mild, moist climate of the Pacific Northwest (coastal forests west of the Cascade crest; Gilliam 2016). A relatively narrow belt of huge coniferous trees – Temperate Rain Forest – extends along the Pacific coastline from northern California to southern Alaska (Chapman and Bolen 2015). Mediterranean forests in moister, often higher locations are the pine and fir dominated forests of California (Safford and Vallejo 2019).

The estimated total area of the temperate coniferous forest biome for the U.S. based on forest-type groups is 1.3 million km^2 (0.5 million mi^2; Table 1.2). Mainly coniferous forests dominated by Douglas fir of the Pacific coast sub-region comprising of Oregon, Washington and California accounted for a total biomass C stock of 3.2 Pg C (3.6 billion tn C) in 2012 (Magerl et al. 2019). Plant biomass C can exceed SOC due to rapid rates of C mineralization but SOC can also be the largest C stock (Adams et al. 2019; De Deyn et al. 2008).

Table 1.2 Forest-type groups of temperate forest biomes in the United States (modified from Johnson and Kern 2003)

Temperate coniferous forest biome	Temperate broadleaf and mixed forest biome
Douglas-fir	Eastern aspen (*Populus*) -birch
Eastern fir-spruce	Western aspen-birch
Hemlock-Sitka spruce	*Chaparral* (*Adenostoma*, *Arctostaphylos*, *Ceonothus*)
Larch (*Larix*)	Elm (*Ulmus*) -ash (*Fraxinus*) -cottonwood (*Populus*)
Loblolly pine (*Pinus taeda* L.) – shortleaf pine (*Pinus echinata* Mill.)	Maple – beech (*Fagus*) -birch
Lodgepole pine	Oak – gum (*Liquidambar*) -cypress (*Taxodium*)
Longleaf pine (*Pinus palustris* Mill.) – slash pine (*Pinus elliottii* Engelm)	Oak – hickory (*Carya*)
Pinyon pine (*Pinus edulis* Engelm.) – juniper (*Juniperus*)	Oak-pine
Ponderosa pine (*Pinus ponderosa* Douglas ex C. Lawson)	Western hardwoods
Redwoods (*Sequoia*)	
Western spruce-fir	
Western white pine (Pinus monticola Douglas ex D. Don)	
White-red-jack pine	

Temperate Broadleaf and Mixed Forest Biome

Hardwoods of widely varying species dominance span most of the eastern U.S. temperate broadleaf and mixed forest biome (abbreviated below as "Temperate broadleaf forest"; Fig. 1.4), except for extreme high elevations and southeastern Coastal Plain pinelands with mixed (conifer/hardwood) forests occurring in parts of the southeast, west and, especially, the Pacific Northwest, of the coastal U.S. (Gilliam 2016). Soils in northern regions are extremely fertile Alfisols while Ultisols of southeastern U.S. are typically acidic and deficient in major nutrients (Chapman and Bolen 2015). Dominant soil orders include also Entisols and Inceptisols, encompassing a broad range of characteristics (Adams et al. 2019). Porous rocky soils are found in the subalpine zone. Poorly developed soils limited in fertility and moisture occur at higher elevations (Chapman and Bolen 2015). Mediterranean forests and woodlands are found mostly on Alfisols and Inceptisols, and sometimes on Entisols (Safford and Vallejo 2019).

The highest tree diversity is found in the mixed mesophytic forest region of the Appalachians with dominant genera including oaks (*Quercus* spp.), maples (*Acer* spp.), hickories (*Carya* spp.) and birch, along with American beech (*Fagus grandifolia*) and yellow poplar (*Liriodendron tulipifera*; Gilliam 2016). *Quercus-Carya* forests occur through secondary succession in the southeastern Coastal Plain. Some of the important genera including fir, juniper, spruce, pine, hemlock and Douglas-fir

Fig. 1.4 Dense stands of hemlocks (*Tsuga* spp.) and northern hardwoods along the Elk Knob Trail, Loyalsock State Forest, Lycoming County, Pennsylvania. (Creative Commons Attribution 2.0 Generic license)

of the montane conifer forests of western U.S. form mixed forests. Hardwoods in those mixed forests include maples, alder (*Alnus* spp.), poplar and mountain ash (*Sorbus* spp.; Gilliam 2016). Commonly, a diverse community of multiple strata of understory trees and shrubs, perennial herbs, and mosses is present (Adams et al. 2019). Woodlands with drought-adapted deciduous and coniferous species (oaks, pines, cypresses) are found under Mediterranean climate in California (Safford and Vallejo 2019).

The estimated total area of the temperate broadleaf forest biome for the U.S. based on forest-type groups is 1.2 million km^2 (0.5 million mi^2; Johnson and Kern 2003).

Tropical Forest Biome

The tropical forest biome includes tropical and subtropical moist broadleaf and coniferous forest biomes (abbreviated in the following as "Tropical moist forest"; Fig. 1.5), and tropical and subtropical dry broadleaf forest biome [abbreviated below as "Tropical dry forest" (TDF)]. The lower-latitude forests of the U.S. fall technically within the subtropical classification but will be referred to as tropical forests in a broader sense in the following sections (Silver et al. 2003). The U.S.-affiliated Atlantic (Caribbean) and Pacific Islands encompass nearly all the U.S. tropical moist forest area while some TDFs are found in Hawaii (Oswalt et al. 2019). Further, subtropical moist forests occur in the southeastern U.S. (Cusack and Marín-Spiotta 2019). Climate of the Holdridge humid tropical forests encompasses 2000 mm (79 in) to >8000 mm (315 in) per year mean annual precipitation (MAP)

Fig. 1.5 Rain Forest of El Yunque, Puerto Rico. (Wikimedia Commons; public domain)

and mean annual temperature (MAT) is ≥18 °C (64 °F; Holdridge et al. 1971). In contrast, several months of severe or absolute drought characterize the climate of TDFs (Mooney et al. 1995).

Soils of tropical moist forests are typically heavily leached and naturally nutrient (i.e., P) deficient from the soil orders Ultisols and Oxisols (Chapman and Bolen 2015; Cusack and Marín-Spiotta 2019). Tropical forests also grow on a wide range of soils that are less weathered and, thus, differ in their nutrient availability. Highly organic soils consisting of rich peat and muck may also be found in more moist regions. Tropical forest soils in Hawaii and Puerto Rico are particularly well-studied in the U.S. (Silver et al. 2003).

Prominent tree species among 349 recorded in Puerto Rico are the naturalized African tulip tree (*Spathodea campanulate*; Marcano-Vega 2017). On the U.S. Virgin Islands and the Commonwealth of the Northern Marina Islands, the tan tan tree (*Leucaena leucocephala*) has the highest biomass storage from a total of 118 and 46 species recorded. Species that account for the highest biomass storage on the other Pacific Islands are Maota (*Dysoxylum* spp.) in American Samoa, lagundi (*Vitex negundo*) on Guam, bkau or apgau (*Rhizophora* spp.) in the Republic of Palau, and ʻōhiʻa lehua (*Metrosideros polymorpha*) in Hawaii (Oswalt et al. 2019).

Tropical forests account for only a small percentage of the U.S. forest area. Estimates for the Caribbean forest area of Puerto Rico range between 3001 km^2 (1159 mi^2) and 4934 km^2 (1905 mi^2; Johnson and Kern 2003; Oswalt et al. 2019)

Fig. 1.6 Shawangunk Grasslands National Wildlife Refuge, New York. (Creative Commons Attribution-Share Alike 3.0)

Forest land is estimated to cover 4040 km^2 (1560 mi^2) in Puerto Rico (U.S. EPA 2021). Forests in the other Caribbean area of the U.S., i.e., the U.S. Virgin Islands cover an estimated 190 km^2 (73 mi^2; Oswalt et al. 2019). Among the U.S. Pacific Islands, forests cover an estimated 158 km^2 (61 mi^2), 283 km^2 (109 mi^2), 414 km^2 (160 mi^2), 244 km^2 (94 mi^2), 581 km^2 (224 mi^2), 94 km^2 (36 mi^2), and 5954 km^2 (2299 mi^2) in American Samoa, Guam, Republic of Palau, Commonwealth of the Northern Mariana Islands, Federated States of Micronesia, Republic of the Marshall Islands, and Hawaii, respectively (Oswalt et al. 2019).

More C in tropical forests is stored in vegetation than in soil as C allocation to aboveground supportive plant organs is high while the tropical climate favors C mineralization (De Deyn et al. 2008).

1.3.2 Grassland and Shrubland Biomes

Temperate Grasslands, Savannas, and Shrublands Biome

The temperate grasslands, savannas, and shrublands biome (abbreviated in the flowing sections as "Temperate grassland") is the natural climax vegetation in regions where rainfall is too low to support the growth of forests (Fig. 1.6; Jones and Donnelly 2004). Grasslands occur where potential evaporation exceeds precipitation, and also where disturbance, management or soil conditions prevent woody growth (Pendall et al. 2018). Grassland is a land-use category on which the plant cover is composed principally of grasses, grass-like plants (i.e., sedges and rushes), forbs, or shrubs suitable for grazing and browsing, and includes both pastures and native rangelands (U.S. EPA 2021). Temperate grasslands are mainly composed of C3 and C4 grasses, and C3 forbs and shrubs (De Deyn et al. 2008). Non-forested grazing lands in the U.S. are differentiated into pastures and rangelands (Paustian et al. 2019). Rangeland can be defined as land on which the indigenous vegetation (climax or natural potential) is predominately grasses, grass-like plants, forbs, or shrubs and is managed as a natural ecosystem (SRM 1998). If rangeland plants are introduced, they are managed similarly. Rangeland includes natural grasslands, savannas, shrublands, many deserts, tundras, alpine communities, marshes and meadows" (SRM 1998). In contrast, pastures are often made up of non-indigenous and/or nonnative species, are often derived from other land covers, and support more intensive and more diverse management practices (e.g., fertilization, irrigation, plant species introduction, grazing management; Paustian et al. 2019). Due to high C allocation belowground especially by extensive fibrous root systems of perennial grasses, and the low proportion of persistent woody structures aboveground, C stocks in aboveground vegetation of temperate grassland are small relative to SOC stocks (De Deyn et al. 2008).

Savannas, deserts, and tundra belong to the land-use category "Grassland" (U.S. EPA 2021). Woody plant communities of low forbs and shrubs, such as mesquite, chaparral, mountain shrub, and pinyon-juniper, are also classified as "Grassland" if they do not meet the criteria for "Forest Land". Sclerophyllous (hard-leaved, with a high weight-to-area ratio) evergreen shrubland with widespread adaptations to intense fire is the chaparral vegetation type in California (Safford and Vallejo 2019). Lower, more xeric, sometimes drought-deciduous shrublands called coastal sage scrub occur also in California. Temperate grasslands include also land managed with agroforestry practices, such as silvopasture and windbreaks, if the land is principally grasses, grass-like plants, forbs, and shrubs suitable for grazing and browsing, and assuming the stand or woodlot does not meet the criteria for "Forest Land" (U.S. EPA 2021).

In the U.S., the western edge of the eastern deciduous forest gives way gradually to a temperate grassland known as the tallgrass prairie or prairie (Chapman and Bolen 2015). This area was characterized originally by rich soils developed on deposits of loess. Mollisols developed in a relatively humid, temperate climate. In contrast, Chernozem soils developed under drier conditions in the western area of the tallgrass prairie. A hard layer of calcium carbonate ('hardpan') formed in these soils at 1.5-m (6 ft) depth. Chestnut soils developed under semiarid conditions while brown soils are characteristic for the driest temperate grassland region. Semiarid means that the ratio of precipitation to potential evapotranspiration is between 0.2 and 0.5. Brown soils are less productive and have a hardpan at about 60-cm (2 ft) depth (Chapman and Bolen 2015)

Indiangrass (*Sorghastrum nutans*), big bluestem (*Andropogon gerardi*), and switchgrass (*Panicum virgatum*) are the renowned species of the tallgrass prairie

(Chapman and Bolen 2015). However, virtually all of the tallgrass prairie is gone. Little bluestem (*Schizachyrium scoparium*), some kinds of grama grass (*Bouteloua* spp.), and various species of needlegrass (*Stipa* spp.) characterize the midgrass prairie in the interior U.S. Farther westward are plains of shorter grasses and a land often affected by drought. Buffalo grass (*Bouteloua dactyloides*) and blue grama (*Bouteloua gracilis*) are the principal species in much of the shortgrass prairie (Chapman and Bolen 2015).

Distinctive temperate grasslands developed also at other locations in the U.S. (Chapman and Bolen 2015). This includes the Palouse Prairie, once covering much of eastern parts of Washington and Oregon and extended into western Idaho. Other distinctive temperate grasslands are annual grasslands in the Central Valley of California, southwestern desert grasslands across the southwestern U.S., grasslands of the Edwards Plateau in south-central Texas, and a thorn scrub-grassland in parts of southern Texas – the Tamaulipan Mezquital (Chapman and Bolen 2015).

The total temperate grassland area in CONUS is about 2.6 million km^2 (1.0 million mi^2; Pendall et al. 2018). The major proportion of temperate grassland area is distributed in the central Great Plains ecoregion, California, and central Florida. The land covers "Shrub, Scrub, Grass, Pasture and Hay" covered together 2.9 million km^2 (1.1 million mi^2) in CONUS (Bliss et al. 2014). From 1982 to 2015, the grassland area in the Great Plains decreased (Liu et al. 2020a), with some of this loss attributed to the expansion of cultivation of biofuel crops (Lark et al. 2015). The total pasture/hay area decreased by 43,477 km^2 (16,787 mi^2) or 7.94% between 2001 and 2016 in CONUS (Homer et al. 2020).

Tundra Biome

The tundra biome (Fig. 1.7) – the coldest of all terrestrial biomes – lies: (i) immediately north of the boreal forest biome tree line at high latitudes (arctic tundra), and (ii) above the timber line and the krummholz zone at high altitudes (alpine tundra; Chapman and Bolen 2015). Soils are often azonal consisting of stony or gravel soils intermixed with clay and silts, lacking a soil profile characterized by discernable horizons. However, mature soils with distinctive profiles may also be found. Permafrost is the dominant feature of arctic tundra substrate (Chapman and Bolen 2015). Near-surface (within 1-m or 39-in soil depth) permafrost is commonly underlying tundra land-cover types in Alaska while organic soils are estimated to underlie a fairly small portion (Zhu and McGuire 2016).

Very low temperatures, short growing seasons, low nutrient (i.e., N) availability and waterlogging are major plant stressors (Du et al. 2020; Prentice et al. 1992). Plants in the heath family (Ericaceae) dominate much of the arctic tundra vegetation including several species from the genus Vaccinium (Chapman and Bolen 2015). The community types shrub tundra, dwarf birch (Betula spp.) heath and cotton grass (*Eriophorum angustifolium*) heath ("tussock heath") can be differentiated. A zone of arctic tundra with permafrost occurs across northern Alaska (Chapman and Bolen 2015). Tundra in Alaska is generally dominated by shrub tundra, followed by graminoid and wetland tundra (Zhu and McGuire 2016). Alpine tundra is found on taller mountains in the U.S. (e.g., peaks in the Rocky Mountains and the Sierra Nevada;

Fig. 1.7 Tundra with shrub willow (*Salix* spp.), Arctic National Wildlife Refuge, Brooks Range, Alaska. (Wikimedia Commons; public domain)

Chapman and Bolen 2015). Here, permafrost is rare or absent, and the flora includes a variety of seed-producing plants and dwarfed species of birch and willow (Salix spp.).

In 2005, tundra land-cover types covered 0.42 million km^2 (0.16 million mi^2) in Alaska (Zhu and McGuire 2016). In comparison, Raynolds et al. (2012) estimated a tundra area of 0.48 million km^2 (0.19 million mi^2). However, large uncertainties regarding tundra extent remain, similarly to that for the boreal forest biome in Alaska (Zhu and McGuire 2016). Further, no data on extent of alpine tundra for CONUS are available.

1.3.3 Wetland Biomes

Terrestrial Wetlands

There are various definitions for terrestrial wetlands which can be described as non-tidal freshwater wetlands (Fig. 1.8; Kolka et al. 2018). However, all definitions recognize a high water table level as the driver of biological and chemical processes characteristic of wetlands. Terrestrial wetlands occur in boreal, temperate, and tropical climate zones in the U.S. The two major soil types in wetlands are mineral soils and organic soils. Wetland ecosystems with organic soils, also known as peatlands, are classified as Histosols (Soil Survey Staff 2010). The Histosol order represents

Fig. 1.8 Mountaintop bog, headwaters of Doe Run, Loyalsock State Forest, Lycoming County, Pennsylvania. (Creative Commons Attribution 2.0)

soils with a thick [>40-cm (16-in)] accumulation of OM on top of mineral sediments or rock. Otherwise, mineral soil wetlands vary widely in the composition and depth of the surface organic layer, varying from a few centimeters (inches) to nearly 40 cm (16 in) in histic-mineral soil wetlands (Soil Survey Staff 2010). The vegetation cover of wetlands is highly variable including non-forested and forested peatlands, and mineral soil wetlands. Fens have vegetation generally dominated by sedges (*Carex* spp.) and brown mosses (Kolka et al. 2018). In contrast, bogs have peat mosses (*Sphagnum* spp.). Other species are found, for example, in bottomland hardwood ecosystems and Atlantic white cedar (*Thuja occidentalis* L.) swamps (Kolka et al. 2018).

Estimates of the extent of terrestrial wetlands for the U.S. are uncertain as these depend partially on definitions used. There may be about 0.40 million km^2 (0.15 million mi^2) of terrestrial wetlands in CONUS (Kolka et al. 2018). Further, the combined area of the land covers "Woody Wetlands" and "Emergent Herbaceous Wetlands" is 0.33 million km^2 (0.13 million mi^2; Bliss et al. 2014). In Alaska alone, however, there may be an estimated 0.58 million km^2 (0.22 million mi^2) of terrestrial wetlands (Kolka et al. 2018). In contrast, Pastick et al. (2017) estimated that wetlands in Alaska cover roughly 177,000 km^2 (68,340 mi^2) of the total land surface area. The peatland areas of CONUS and Alaska have been estimated at 0.11 million km^2 (0.04 million mi^2) and 0.13 million km^2 (0.05 million mi^2), respectively (Bridgham et al. 2000, 2006). Recently refined estimates of peatland area for the U.S. range between 0.20 and 0.25 million km^2 (0.08–0.10 million mi^2; Xu et al. 2018).

1.3.4 Desert and Xeric Shrubland Biome

Deserts and xeric shrublands (abbreviated as "Deserts") in the U.S. include the Great Basin shrub steppe and Wyoming Basin shrub steppe in colder regions, and in warmer regions the Colorado Plateau shrublands, Mojave desert and Snake–Columbia shrub steppe, and parts of the Chihuahuan desert, Sonoran desert and Tamaulipan mezquital (Fig. 1.9; Chapman and Bolen 2015). Deserts are sometimes also classified as grasslands (Cobon et al. 2017). Among the soils of the Desert Biome are Aridisols formed under the influence of high temperatures, low but often torrential precipitation, and wind (Chapman and Bolen 2015). Deeper within the profile of desert soils materials may be cemented into 'caliche', a hardpan of calcium carbonate. Lithosols, i.e., undeveloped strata composed mainly of rocks are common in desert areas subject to heavy erosion. Further, Mollisols may be found in the Great Basin while dunes may occur at other sites. Hardened surfaces, i.e., desert pavement, and crust may also develop on soil surfaces in arid regions (Chapman and Bolen 2015).

Among plant species in desert ecosystems are phreatophytes, plants whose deep root systems penetrate into permanent sources of water (Chapman and Bolen 2015). Examples for phreatophytes are willows (*Salix* spp.) and salt cedar (*Tamarix* spp.). Among other plant species are desert-willow (*Chilopsis linearis*), smoke tree

Fig. 1.9 Mesquite (*Prosopis* spp.) bosque, Desert National Wildlife Refuge, Clark County, Nevada. (Creative Commons Zero, Public Domain Dedication)

(*Cotinus* spp.), and large mesquite. Salt-tolerant plants known as halophytes are also found, i.e., various species of glasswort (*Salicornia* spp.). A high diversity of ephemerals, plants that temporarily appear on an irregular basis rather than each year grow in deserts and shrublands. Drought-enduring plants such as cacti, the desert tree palo verde (*Parkinsonia* spp.) and the shrub ocotillo (*Fouquieria splendens*) are also found. Key species in the Chihuahuan Desert include creosote bush (*Larrea tridentate*), tarbush (*Flourensia cernua*), sotol (*Dasylirion* spp.), ocotillo, and several yuccas and agaves. The agave lechuguilla (*Agave lechuguilla*) is an indicator species. A familiar indicator for much of the Sonoran Desert is the cactus giant saguaro (*Carnegiea gigantea*). The Mojave Desert lacks plant species diversity with creosote bush often covering large areas. The yucca joshua tree (Yucca brevifolia) is the most distinctive plant. Species of sagebrush, saltbush (*Atriplex* spp.), rabbitbrush (*Chrysothamnus* spp.), and greasewood (common name shared by several plants) are among the dominant plants in the Great Basin Desert. Major grasses include blue bunch wheatgrass (*Pseudoroegneria spicata*) and squirrel tail (*Elymus elymoides*), but several other species are also important such as Indian rice grass (*Oryzopsis hymenoides*). However, the Eurasian cheatgrass (*Bromus tectorum*) has invaded much of the Great Basin (Chapman and Bolen 2015).

The areal extent of the desert biome in the U.S. is uncertain as, for example, some desert areas are considered rangeland (United States Department of Agriculture 2018). Further, arid desert grasslands and shortgrass steppe may also be classified under the desert biome (Pendall et al. 2018). The land covers "Barren" and "Shrub/Scrub" are estimated to cover about 54,000 km^2 (21,000 mi^2) and 1.3 million km^2 (0.5 million mi^2), respectively (Bliss et al. 2014). Between 2001 and 2016, the total shrub extent in CONUS increased by 4512 km^2 (1742 mi^2) or 0.26% (Homer et al. 2020). During the same period, barren land area increased by 1077 km^2 (416 mi^2) or 1.32% (Homer et al. 2020).

1.3.5 Anthropogenic Biome Group Cropland

Humans have shaped most of terrestrial nature for at least 12,000 years (Fig. 1.10; Ellis et al. 2021). However, existing descriptions of biome systems either ignore human influence altogether or describe it using at most four anthropogenic ecosystem classes (urban/built-up, cropland, and one or two cropland/natural vegetation mosaic(s); Ellis and Ramankutty 2008). Ellis and Ramankutty (2008) presented the first characterization of terrestrial biomes based on global patterns of sustained, direct human interaction with ecosystems. Eighteen "anthropogenic biomes" were identified through empirical analysis of global population, land use, and land cover. The anthropogenic biome group croplands are mostly mosaics of cultivated land mixed with trees and pastures. Thus, this biome group constitutes about half of the world's total crop-covered area (Ellis and Ramankutty 2008).

Cropland is a land-use category that includes areas used for producing adapted crops for harvest (U.S. EPA 2021). This category includes both cultivated and

Fig. 1.10 Corn (*Zea mays* L.) field, Moore Township, Northampton County, Pennsylvania. (Creative Commons Attribution 2.0)

non-cultivated lands. The total cropland area in CONUS is considered managed land area (Ogle et al. 2018). Managed croplands are found mostly in areas accessible to human activity, and include altering or maintaining the condition of the land to produce commercial or non-commercial products or services (U.S. EPA 2021). Cultivated crops include row crops or close-grown crops, and hay or pasture in rotation with cultivated crops. Non-cultivated cropland includes continuous hay, perennial crops (e.g., orchards) and horticultural cropland. Cropland also includes land with agroforestry, such as alley cropping and windbreaks, if the dominant use is crop production, assuming the stand or woodlot does not meet the criteria for "Forest Land". Lands in temporary fallow or enrolled in conservation reserve programs (i.e., set-asides) are also classified as "Cropland", as long as these areas do not meet the "Forest Land" criteria (U.S. EPA 2021).

The major crops planted in U.S. croplands are corn, wheat, soybean, and cotton (*Gossypium* spp.; Hristov et al. 2018). Another major cropland product is hay. In 2017, 275,000, 273,000, 157,000, 116,000 and 38,000 km^2 (106,178, 105,406, 60,618, 44,788 and 14,672 mi^2) were under corn, soybean, hay, wheat and cotton, respectively (Crossley et al. 2021). Typically, annual cropping systems are managed intensively including tillage, crop species and crop rotation, residue management, fertilizer and nutrient inputs, extent and efficiency of drainage, and irrigation, and use of cover crops. Perennial cropland systems avoid the 4- to 8-month fallow period common among many annual row-crop systems. Aside grasslands, pasture, and hayed lands, perennial crops include tree crops (i.e., fruits, nuts) and vineyards.

Many perennial fruit, nut, and vegetable systems generally are intensively managed. Types of management include cover crops and intercropping, irrigation and tillage, fertilizer use, and cultural activities. Biofuel feedstock crops, including perennial grasses and short-rotation woody crops, are also under cultivation in U.S. croplands (Hristov et al. 2018).

About 0.8 million km^2 (0.3 million mi^2) of arable land in the U.S. is affected by a single land degradation process with the dominant pathway aridity (Prăvălie et al. 2021). Major land degradation processes considered in this assessment included aridity, vegetation decline, soil erosion, soil salinization and SOC decline. Land degradation involves a negative trend in land (soils, vegetation and water resources) condition. Globally, the U.S. is among the main hotspots, in terms of absolute data, of both uni-degradation and multi-degradation in arable lands (Prăvălie et al. 2021).

Cropland, i.e., all land considered to be in crop rotation (cropland used for crops, idle cropland, and cropland used only for pasture) in CONUS, Alaska and Hawaii covered in total an estimated 1.6 million km^2 (0.6 million mi^2) in the year 2012 (Bigelow and Borchers 2017). However, a similar extent was also reported for the cropland area of CONUS alone (Hristov et al. 2018). These numbers are affected greatly by the inclusion or exclusion of Alaska, which, relative to the rest of the U.S., has small amounts of cropland (Bigelow and Borchers 2017). Otherwise, Bliss et al. (2014) estimated that the land cover row crops covered 1.2 million km^2 (0.5 million mi^2). Specifically, corn, soybean, wheat, cotton, and hay covered an estimated 0.4 million km^2 (0.1 million mi^2), 0.3 million km^2 (0.1 million mi^2), 0.2 million km^2 (0.08 million mi^2), 0.04 million km^2 (0.01 million mi^2), and 0.2 million km^2 (0.09 million mi^2), respectively (Hristov et al. 2018). Cropped organic soils such as Histosols comprise <1% of total cropland area in CONUS (Hristov et al. 2018). From 1982 to 2015, the cropland area in the Great Plains increased (Liu et al. 2020a). In CONUS, cultivated crop area gained 48,555 km^2 (18,747 mi^2) between 2001 and 2016, largely at the expense of the area under pasture/hay (Homer et al. 2020).

1.3.6 Anthropogenic Biome Group Dense Settlements and Villages

Most people live in cities and towns embedded within the anthropogenic biome groups dense settlements and villages (Fig. 1.11; Ellis and Ramankutty 2008). Urban, dense settlements, human settlements and villages are considered "urban areas" in the following sections and chapters. Urban areas are affected by various levels of human activities with artificial impervious cover strongly altering vegetation and biogeochemical cycles and, thus, SOC dynamics (Lorenz and Lal 2017). In 2018, the artificial impervious cover in the U.S. was >40,000 km^2 (15,500 mi^2; Gong et al. 2020).

There is no single definition of "urban", and objective methods to spatially enclose urban areas and underlying data are also uncertain (Gurney et al. 2018). For

Fig. 1.11 Lower Manhattan, New York City, New York. (Creative Commons Attribution-Share Alike 4.0)

example, the NRI category of developed land includes: (i) large tracts of urban and built-up land; (II) small tracts of built-up land of less than 4 ha (10 acres); and (III) land outside of these built-up areas that is in a rural transportation corridor (roads, railroads, and associated rights-of-way; USDA 2018). However, NRI does not inventory Federal lands (Nelson et al. 2020).

Urban areas and dense settlements can also be differentiated by population densities with urban areas having population densities of 3172 persons km^{-2} (8215 persons mi^{-2}), and population densities of 807 persons km^{-2} (2090 persons mi^{-2}) for dense settlements, respectively (Ellis and Ramankutty 2008). In contrast, the U.S. Census Bureau defines urban lands by population densities of at least 386 people km^{-2} (1000 people mi^{-2}) while surrounding or exurban areas have densities of 193 people km^{-2} (500 people mi^{-2}; https://www.census.gov/). Census Bureau definitions are not based on land use or land cover, but instead measure human population density (Nelson et al. 2020). This lack of consensus also affects credibility of data on the extent of urban land cover (Table 1.3). Nevertheless, U.S. urban areas have increased in extent. For example, from 1973 to 2000, the developed land area in CONUS increased from 235,000 km^2 (90,734 mi^2) to 313,000 km^2 (120,850 mi^2; Sleeter et al. 2013). Development area includes residential, industrial, commercial, transportation, and areas such as parks or other open spaces surrounded or otherwise dominated by an urban landscape. Based on urban boundaries delineated using 30 m artificial impervious area, Li et al. (2020) estimated a total urban boundary area for the U.S. of 153,294 km^2 (59,187 mi^2) and 218,073 km^2

Table 1.3 Extent of "urban" area for different regions in the United States

Region	Extent		Reference
	km^2	mi^2	
U.S. urban land	802,054	309,675	CIESIN (2013)
	275,538	106,386	www.census.gov
Urban area conterminous U.S., Alaska and Hawaii	283,280	109,375	Bigelow and Borchers (2017)
Urban and built-up area conterminous U.S., Hawaii, Puerto Rico and Virgin Islands	375,548	145,000	USDA (2018)
Developed land area conterminous U.S.	383,375	148,022	Bliss et al. (2014)
Urban anthropogenic biome conterminous U.S.	216,720	83,676	Ellis and Ramankutty (2008)
Dense settlements anthropogenic biome conterminous U.S.	247,904	95,716	
U.S. urban boundary area	218,073	84,198	Li et al. (2020)

(84,198 mi^2) in 2000 and 2018, respectively. The total developed area in CONUS reached 428,575 km^2 (165,474 mi^2) in 2016, which was a net increase of 28,626 km^2 (11,053 mi^2) or 6.7% compared to that in 2001 (Homer et al. 2020).

1.4 Changes in the Terrestrial Biosphere by Human Activities

Similar to many other global regions, humans have restructured the terrestrial biosphere of the U.S. for agriculture, forestry, and other, i.e., urban uses. For example, urban expansion occupied 66,400 km^2 (25,600 mi^2) of the U.S. land area from 1992 to 2016 with 16,000 km^2 (6200 mi^2) of cropland converted corresponding to a loss of 6.72 Tg C (7.4 billion tn C) cropland NPP (Huang et al. 2020). From 2001 to 2016 alone, 44,500 km^2 (17,200 mi^2) of CONUS farm- and ranchland were paved over, fragmented, or converted to uses that compromise agriculture (Freedgood et al. 2020). This area is equal to the total amount of cropland of Ohio. The total farmland loss included 18,000 km^2 (6900 mi^2) of the most productive, versatile, and resilient land in the U.S. Within the urban boundary of the U.S., the impervious surface area increased from 86,613 km^2 (33,441 mi^2) in 2000 to 130,863 km^2 (50,526 mi^2) in 2018 (Li et al. 2020). Some of the lost farmland and the newly covered urban land may have offered a high potential for C sequestration (Freedgood et al. 2020). Historically, urban areas in CONUS increased from 0.8% of the national share of land area in 1945 to 3.7% in 2012 (Spangler et al. 2020). However, expansion of urban areas in the U.S. may now be decelerating (Liu et al. 2020b). From "peak expansion", urban expansion in the continental U.S. may have already decreased by 10,000 km^2 (3,860 mi^2; Richter 2020).

In 1945, cropland use occurred on 23.7% of the national share of land use in CONUS, and on 20.7% in 2012 (Spangler et al. 2020). This slight decrease occurred

Table 1.4 Surface area of anthropogenic biomes in the United States and percentage of total United States land area (8.97 million km^2 [3.46 million mi^2]) covered by each anthropogenic biome in the year 2000 (Erle C. Ellis, personal communication)

Anthropogenic biome	Area		
	km^2	mi^2	%
Urban	216,720	83,676	2.4
Dense settlements	247,904	95,716	2.8
Irrigated villages	77	30	0.0
Pastoral villages	76	29	0.0
Residential irrigated croplands	91,688	35,401	1.0
Residential rainfed mosaic croplands	837,281	323,276	9.3
Populated irrigated croplands	1,132,354	437,204	12.6
Populated rainfed croplands	755,242	291,601	8.4
Residential rangelands	140,853	54,384	1.6
Populated rangelands	442,227	170,745	4.9
Remote rangelands	1,617,406	624,484	18.0
Populated forests	1,241,335	479,282	13.9
Remote forests	363,321	191,019	4.1
Wild forests	1,312,242	506,659	14.6
Sparse trees	571,554	220,678	6.4

primarily within the past four decades. From 2008 to 2016, 0.041 million km^2 (0.016 million mi^2) of land were converted to cropland with gross expansion minus gross abandonment area of 1538 to 5382 km^2 (594 to 2078 mi^2; Lark et al. 2020). Grasslands, including those used for pasture and hay, constituted 88% of the land converted to crop production across CONUS. In comparison, special-use areas increased from 4.5% in 1945 to 8.9% in 2012 (Spangler et al. 2020). Special use areas include rural transportation, rural parks and wildlife, defense and industrial areas, and miscellaneous farmland (farmsteads, farm roads and lanes, and miscellaneous farmland). Grassland, pasture and range land decreased by 0.03% in 1945–2012. Further, forest-use decreased from 31.6% in 1945 to 28.5% in 2012, and miscellaneous land uses decreased in the same period from 4.9% to 3.6% in CONUS (Spangler et al. 2020).

The patterns of species composition and abundance, primary productivity, land-surface hydrology, and the biogeochemical cycles of C, N, and P, have all been substantially altered by human activities globally (Ellis and Ramankutty 2008). Managed land, for example, accounted for up to 45% of the interannual variability of the net land C balance over 1959–2015 (Yue et al. 2020). In addition to human land use, natural disturbances are key factors in the C balances of terrestrial ecosystems of CONUS (Liu et al. 2020b). Human land use specifically affects the potential for SOC sequestration by the U.S. biosphere. Seventeen of the anthropogenic biomes introduced by Ellis and Ramankutty (2008) are also present in the terrestrial U.S. biosphere (Table 1.4).

In the year 2000, the largest anthropogenic biomes in the U.S. were remote rangelands described as rangelands used for livestock grazing and covered by minimal crops and forests, and with inconsequential human populations (Ellis and Ramankutty 2008). However, used land areas in the mapping of anthropogenic biomes, and of rangelands in particular, may be overestimated (Ellis et al. 2020). The second largest anthropogenic biomes in the U.S. were wild forests which are wildlands without human populations or agriculture characterized by high tree cover, mostly consisting of boreal and tropical forests (Ellis and Ramankutty 2008). The third largest U.S. anthropogenic biome area was covered by populated forests with minor human populations followed by populated irrigated croplands which are described as irrigated croplands with annual crops mixed with other land uses and land cover, and also with minor human populations. All other anthropogenic biomes covered less than 10% each of the total U.S. land area in 2000 with the largest being residential rainfed mosaic croplands (mix of trees and rainfed cropland with substantial human populations), populated rainfed croplands (rainfed cropland with minor human populations), and remote forests (forests with inconsequential human populations; Ellis and Ramankutty 2008).

1.5 Conclusions

The terrestrial biosphere of the U.S. can be divided into biomes which contain multiple terrestrial ecosystems characterized by common physical properties. The temperate grassland, savanna, and shrubland biome is the largest terrestrial biome in the U.S. followed by the 'forest biome', i.e., the combined temperate coniferous, broadleaf and mixed, and tropical forests biomes. Large areas in the U.S. are also under cropland use. Strongly increasing in extent in CONUS are urban areas. Terrestrial wetlands and peatlands, some of them underlain by permafrost contain important SOC stocks due to unfavorable environmental conditions for decomposition. Humans have restructured the terrestrial biosphere of the U.S. for agriculture, forestry, urban and other uses. Specifically, the area under cropland was recently expanded mainly at the expense of the grassland area. Land management accounts for about half of the interannual variability in the net land C balance with potential effects on SOC sequestration by terrestrial biomes in the U.S.

References

Adams MB, Kelly C, Kabrick J, Schuler J (2019) Temperate forests and soils. In: Busse M, Giardina CP, Morris DM, Page-Dumroese DS (eds) Global change and forest soils, Developments in soil science 36. Elsevier, Amsterdam, pp 83–108. https://doi.org/10.1016/B978-0-444-63998-1.00006-9

Allison SD, Treseder K (2011) Climate feedbacks to microbial decomposition in boreal soils. Fungal Ecol 4:362–374. https://doi.org/10.1016/j.funeco.2011.01.003

Barrett S, Dasgupta A, Dasgupta P et al (2020) Social dimensions of fertility behavior and consumption patterns in the Anthropocene. Proc Natl Acad Sci USA 117:6300–6307. https://www.pnas.org/cgi/doi/10.1073/pnas.1909857117

Bernier P, Paré D, Stinson G et al (2017) Moving beyond the concept of "primary forest" as a metric of forest environment quality. Ecol Appl 27:349–354. https://doi.org/10.1002/eap.1477

Bigelow DP, Borchers A (2017) Major uses of land in the United States, 2012. EIB-178. U.S. Department of Agriculture, Economic Research Service

Birdsey R, Pregitzer K, Lucier A (2006) Forest carbon management in the United States: 1600–2100. J Environ Qual 35:1461–1469. https://doi.org/10.2134/jeq2005.0162

Birdsey R, Mayes MA, Romero-Lankao P et al (2018) Executive summary. In: Cavallaro N, Shrestha G, Birdsey R et al (eds) Second state of the carbon cycle report (SOCCR2): a sustained assessment report. U.S. global change research program, Washington, DC, pp 21–40. https://doi.org/10.7930/SOCCR2.2018.ES

Bliss NB, Waltman SW, West LT, Neale A, Mehaffey M (2014) Distribution of soil organic carbon in the conterminous United States. In: Hartemink A, McSweeney K (eds) Soil carbon. Progress in soil science. Springer, Cham, pp 85–93. https://doi.org/10.1007/978-3-319-04084-4_9

Bonan GB (2008) Forests and climate change: forcings, feedbacks, and the climate benefits of forests. Science 320:1444–1449. https://doi.org/10.1126/science.1155121

Bridgham SD, Ping CL, Richardson L, Updegraff K (2000) Soils of northern peatlands: histosols and gelisols. In: Richardson JL, Vepraskas MJ (eds) Wetland soils: genesis, hydrology, landscapes, and classification. CRC Press, Boca Raton, pp 343–370

Bridgham SD, Megonigal JP, Keller JK, Bliss NB, Trettin C (2006) The carbon balance of North American wetlands. Wetlands 26:889–916

Campbell NA, Reece JB (2005) Biology, 7th edn. Pearson, Benjamin Cummings, San Francisco

Carpenter DN, Bockheim JG, Reich PF (2014) Soils of temperate rainforests of the North American Pacific Coast. Geoderma 230(231):250–264. https://doi.org/10.1016/j.geoderma.2014.04.023

Cavicchioli R, Ripple WJ, Timmins KN et al (2019) Scientists' warning to humanity: microorganisms and climate change. Nat Rev Microbiol 17:569–586. https://doi.org/10.1038/s41579-019-0222-5

Center for International Earth Science Information Network (CIESIN)/Columbia University (2013) Urban-rural population and land area estimates version 2. NASA Socioeconomic Data and Applications Center (SEDAC), Palisades. https://data.worldbank.org/indicator/AG.LND.TOTL.UR.K2

Central Intelligence Agency (2020) The world factbook. https://www.cia.gov/library/publications/the-world-factbook/geos/us.html

Chapman BR, Bolen EG (2015) Ecology of North America, 2nd edn. Wiley, West Sussex

Cobon DH, Baethgen WE, Landman W et al (2017) Agroclimatology in grasslands. In: Hatfield JL, Sivakumar MVK, Prueger JH (eds) Agroclimatology: linking agriculture to climate. Agronomy Monograph 60. ASA, CSSA, and SSSA, Madison. https://doi.org/10.2134/agronmonogr60.2016.0013

Conradi T, Slingsby JA, Midgley GF et al (2020) An operational definition of the biome for global change research. New Phytol 227:1294–1306

Crossley MS, Burke KD, Schoville SD, Radeloff VC (2021) Recent collapse of crop belts and declining diversity of US agriculture since 1840. Glob Chang Biol 27:151–164. https://doi.org/10.1111/gcb.15396

Cusack DF, Marín-Spiotta E (2019) Wet tropical soils and global change. In: Busse M, Giardina CP, Morris DM, Page-Dumroese DS (eds) Global change and forest soils, Developments in soil science 36. Elsevier, Amsterdam, pp 131–169. https://doi.org/10.1016/B978-0-444-63998-1.00008-2

Dahl TE (1990) Wetlands losses in the United States: 1780's to 1980's. U.S. Department of the Interior, Fish and Wildlife Service, Washington, DC, 13p. http://www.fws.gov/wetlands/Status-And-Trends/index.html

De Deyn CB, Cornelissen JHC, Bardgett RD (2008) Plant functional traits and soil carbon sequestration in contrasting biomes. Ecol Lett 11:516–531. https://doi.org/10.1111/j.1461-0248.2008.01164.x

DeLuca TH, Boisvenue C (2012) Boreal forest soil carbon: distribution, function and modelling. Forestry 85:161–184. https://doi.org/10.1093/forestry/cps003
Du E, Terrer C, Pellegrini AFA et al (2020) Global patterns of terrestrial nitrogen and phosphorus limitation. Nat Geosci 13:221–226. https://doi.org/10.1038/s41561-019-0530-4
Ellis EC, Ramankutty N (2008) Putting people in the map: anthropogenic biomes of the world. Front Ecol Environ 6:439–447. https://doi.org/10.1890/070062
Ellis EC, Beusen AHW, Goldewijk KK (2020) Anthropogenic biomes: 10,000 BCE to 2015 CE. Land 9:129. https://doi.org/10.3390/land9050129
Ellis EC, Gauthier N, Goldewijk KK et al (2021) People have shaped most of terrestrial nature for at least 12,000 years. Proc Natl Acad Sci USA 118:e2023483118. https://doi.org/10.1073/pnas.2023483118
Finlay RD, Mahmood S, Rosenstock N et al (2020) Reviews and syntheses: biological weathering and its consequences at different spatial levels – from nanoscale to global scale. Biogeosciences 17:1507–1533. https://doi.org/10.5194/bg-17-1507-2020
Food and Agriculture Organization of the United Nations FAO (2006) Global forest resources assessment 2005. FAO Forestry Department, Forest Resources Assessment Programme, FAO Forestry Paper 147. http://www.fao.org/docrep/008/a0400e/a0400e00.htm
Freedgood J, Hunter M, Dempsey J, Sorensen A (2020) Farms under threat: the state of the states. American Farmland Trust, Washington, DC
Gilliam FS (2016) Forest ecosystems of temperate climatic regions: from ancient use to climate change. New Phytol 212:871–887. https://doi.org/10.1111/nph.14255
Gong P, Li X, Wang J et al (2020) Annual maps of global artificial impervious area (GAIA) between 1985 and 2018. Remote Sens Environ 236:111510. https://doi.org/10.1016/j.rse.2019.111510
Guo Y, Gong P, Amundson R, Yu Q (2006) Analysis of factors controlling soil carbon in the conterminous United States. Soil Sci Soc Am J 70:601–612. https://doi.org/10.2136/sssaj2005.0163
Gurney KR, Romero-Lankao P, Pincetl S et al (2018) Chapter 4: Understanding urban carbon fluxes. In: Cavallaro N, Shrestha G, Birdsey R et al (eds) Second state of the carbon cycle report (SOCCR2): a sustained assessment report. U.S. global change research program, Washington, DC, pp 189–228. https://doi.org/10.7930/SOCCR2.2018.Ch4
Hayes DJ, Turner DP, Stinson G et al (2012) Reconciling estimates of the contemporary North American carbon balance among terrestrial biosphere models, atmospheric inversions, and a new approach for estimating net ecosystem exchange from inventory-based data. Glob Chang Biol 18:1282–1299. https://doi.org/10.1111/j.1365-2486.2011.02627.x
Holdridge L, Grenke W, Hatheway W, Liang T, Tosi J (1971) Forest environments in tropical life zones. Pergamon Press, New York
Homer C, Dewitz J, Jin S et al (2020) Conterminous United States land cover change patterns 2001–2016 from the 2016 National Land Cover Database. ISPRS J Photogramm 162:184–199. https://doi.org/10.1016/j.isprsjprs.2020.02.019
Hristov AN, Johnson JMF, Rice CW et al (2018) Chapter 5: Agriculture. In: Cavallaro N, Shrestha G, Birdsey R et al (eds) Second state of the carbon cycle report (SOCCR2): a sustained assessment report. U.S. global change research program, Washington, DC, pp 229–263. https://doi.org/10.7930/SOCCR2.2018.Ch5
Huang Q, Liu Z, He C et al (2020) The occupation of cropland by global urban expansion from 1992 to 2016 and its implications. Environ Res Lett 15:084037. https://doi.org/10.1088/1748-9326/ab858c
Johnson MG, Kern JS (2003) Quantifying the organic carbon held in forested soils of the United States and Puerto Rico. In: Kimble JS, Lal R, Birdsey R, Heath LS (eds) The potential of U.S. forest soils to sequester carbon and mitigate the greenhouse effect. CRC Press LLC, Boca Raton, pp 47–72
Jones MB, Donnelly A (2004) Carbon sequestration in temperate grassland ecosystems and the influence of management, climate and elevated CO_2. New Phytol 164:423–439. https://doi.org/10.1111/j.1469-8137.2004.01201.x
Kolka R, Trettin C, Tang W et al (2018) Chapter 13: Terrestrial wetlands. In: Cavallaro N, Shrestha G, Birdsey R et al (eds) Second state of the carbon cycle report (SOCCR2): a sustained assess-

ment report. U.S. global change research program, Washington, DC, pp 507–567. https://doi.org/10.7930/SOCCR2.2018.Ch13

Kranabetter JM, Sholinder A, de Montingy L (2020) Contrasting conifer species productivity in relation to soil carbon, nitrogen and phosphorus stoichiometry of British Columbia perhumid rainforests. Biogeosciences 17:1247–1260. https://doi.org/10.5194/bg-17-1247-2020

Lark TJ, Salmon JM, Gibbs HK (2015) Cropland expansion outpaces agricultural and biofuel policies in the United States. Environ Res Lett 10:044003. https://doi.org/10.1088/1748-9326/10/4/044003

Lark TJ, Spawn SA, Bougie M, Gibbs HK (2020) Cropland expansion in the United States produces marginal yields at high costs to wildlife. Nat Commun 11:4295. https://doi.org/10.1038/s41467-020-18045-z

Li X, Gong P, Zhou Y et al (2020) Mapping global urban boundaries from the global artificial impervious area (GAIA) data. Environ Res Lett 15:094044. https://doi.org/10.1088/1748-9326/ab9be3

Liu H, Gong P, Wang J et al (2020a) Annual dynamics of global land cover and its long-term changes from 1982 to 2015. Earth Syst Sci Data 12:1217–1243. https://doi.org/10.5194/essd-12-1217-2020

Liu J, Sleeter BM, Zhu Z et al (2020b) Critical land change information enhances the understanding of carbon balance in the United States. Glob Chang Biol 26:3920–3929. https://doi.org/10.1111/gcb.15079

Liu N, Caldwell PV, Dobbs GR et al (2021) Forested lands dominate drinking water supply in the conterminous United States. Environ Res Lett 16:084008. https://doi.org/10.1088/1748-9326/ac09b0

Lorenz K, Lal R (2017) Impacts of land take and soil sealing on soil carbon. In: Gardi C (ed) Urban expansion, land cover and soil ecosystem services. Taylor & Francis, Routledge, London, pp 181–192

Magerl A, Le Noë J, Erb KH, Bhan M, Gingrich S (2019) A comprehensive data-based assessment of forest ecosystem carbon stocks in the US 1907–2012. Environ Res Lett 14:125015. https://doi.org/10.1088/1748-9326/ab5cb6

Marcano-Vega H (2017) Forests of Puerto Rico, 2014. Resour. Update FS–121. U.S. Department of Agriculture, Forest Service, Southern Research Station, Asheville

Merrill D, Leatherby L (2018) Here's how America uses its land. Bloomberg Press. https://www.bloomberg.com/graphics/2018-us-land-use/

Mooney HA, Bullock SH, Medina E (1995) Introduction. In: Bullock SH, Mooney HA, Medina E (eds) Seasonally dry tropical forests. Cambridge University Press, Cambridge, pp 1–8

Mucina L (2019) Biome: evolution of a crucial ecological and biogeographical concept. New Phytol 222:97–114. https://doi.org/10.1111/nph.15609

Nelson MD, Riitters KH, Coulston JW et al (2020) Defining the United States land base: a technical document supporting the USDA Forest Service 2020 RPA assessment. Gen Tech Rep NRS-191. U.S. Department of Agriculture, Forest Service, Northern Research Station, Madison. 70 p. https://doi.org/10.2737/NRS-GTR-191

Ogle SM, Domke G, Kurz WA et al (2018) Delineating managed land for reporting national greenhouse gas emissions and removals to the United Nations framework convention on climate change. Carbon Balance Manage 13:9. https://doi.org/10.1186/s13021-018-0095-3

Oswalt SN, Smith WB, Miles PD et al (2019) Forest resources of the United States, 2017. A technical document supporting the Forest Service 2020 RPA Assessment, Gen Tech Rep WO-97. U.S. Department of Agriculture Forest Service, Washington Office, Washington, DC, p 223. https://doi.org/10.2737/WO-GTR-97

Pastick NJ, Duffy P, Genet H et al (2017) Historical and projected trends in landscape drivers affecting carbon dynamics in Alaska. Ecol Appl 27:1383–1402

Pattison R, Andersen HE, Gray A et al (eds) (2018) Forests of the Tanana Valley State Forest and Tetlin National Wildlife Refuge Alaska: results of the 2014 pilot inventory, Gen Tech Rep PNW-GTR 967, vol 967. U.S. Department of Agriculture Forest Service, Pacific Northwest Research Station, Portland

Paustian K, Larson E, Kent J, Marx E, Swan A (2019) Soil C sequestration as a biological negative emission strategy. Front Clim 1:8. https://doi.org/10.3389/fclim.2019.00008

Pendall E, Bachelet D, Conant RT et al (2018) Chapter 10: Grasslands. In: Cavallaro N, Shrestha G, Birdsey R et al (eds) Second state of the carbon cycle report (SOCCR2): a sustained assessment report. U.S. global change research program, Washington, DC, pp 399–427. https://doi.org/10.7930/SOCCR2.2018.Ch10

Prăvălie R, Patriche C, Borrelli P et al (2021) Arable lands under the pressure of multiple land degradation processes. A global perspective. Environ Res 194:110697. https://doi.org/10.1016/j.envres.2020.110697

Prentice IC, Cramer W, Harisson SP et al (1992) A global biome model based on plant physiology and dominance, soil properties and climate. J Biogeogr 19:117–134

Raynolds MK, Walker DA, Epstein HE, Pinzon JE, Tucker CJ (2012) A new estimate of tundra-biome phytomass from trans-Arctic field data and AVHRR NDVI. Remote Sens Lett 3:403–411. https://doi.org/10.1080/01431161.2011.609188

Richter MS (2020) Revisiting urban expansion in the continental United States. Landsc Urban Plan 204:103911. https://doi.org/10.1016/j.landurbplan.2020.103911

Safford HD, Vallejo VR (2019) Ecosystem management and ecological restoration in the Anthropocene: integrating global change, soils, and disturbance in boreal and Mediterranean forests. In: Busse M, Giardina CP, Morris DM, Page-Dumroese DS (eds) Global change and forest soils, Developments in soil science 36. Elsevier, Amsterdam, pp 259–306. https://doi.org/10.1016/B978-0-444-63998-1.00012-4

Schimper AFW (1898) Pflanzengeographie auf physiologischer Grundlage. Gustav Fischer Verlag, Jena

Schuur EAG, McGuire AD, Romanovsky V, Schädel C, Mack M (2018) Chapter 11: Arctic and boreal carbon. In: Cavallaro N, Shrestha G, Birdsey R et al (eds) Second state of the carbon cycle report (SOCCR2): a sustained assessment report. U.S. global change research program, Washington, DC, pp 428–468. https://doi.org/10.7930/SOCCR2.2018.Ch11

Silver WL, Lugo AE, Farmer D (2003) Soil organic carbon in tropical forests of the United States of America. In: Kimble JS, Lal R, Birdsey R, Heath LS (eds) The potential of U.S. forest soils to sequester carbon and mitigate the greenhouse effect. CRC Press LLC, Boca Raton, pp 363–382

Sleeter BM, Sohl TL, Loveland TR et al (2013) Land-cover change in the conterminous United States from 1973 to 2000. Glob Environ Chang 23:733–748. https://doi.org/10.1016/j.gloenvcha.2013.03.006

Society for Range Management [SRM] Glossary Update Task Group (1998) Glossary of terms used in range management, 4th edn. Available at https://globalrangelands.org/glossary/

Soil Survey Staff (2010) Key to soil taxonomy, 11th edn. U.S. Department of Agriculture Natural Resource Conservation Service

Song XP, Hansen MC, Stehman SV et al (2018) Global land change from 1982 to 2016. Nature 560:639–643. https://doi.org/10.1038/s41586-018-0411-9

Spangler K, Burchfield EK, Schumacher B (2020) Past and current dynamics of U.S. agricultural land use and policy. Front Sustain Food Syst 4(98). https://doi.org/10.3389/fsufs.2020.00098

Thiffault E (2019) Boreal forests and soils. In: Busse M, Giardina CP, Morris DM, Page-Dumroese DS (eds) Global change and forest soils, Developments in soil science 36. Elsevier, Amsterdam, pp 59–82. https://doi.org/10.1016/B978-0-444-63998-1.00005-7

United States Census Bureau (2012) 2010 Census of population and housing, population and housing unit counts, CPH-2-1, United States summary. U.S. Government Printing Office, Washington, DC

United States Department of Agriculture (2018) Summary report: 2015 national resources inventory. Natural Resources Conservation Service/Center for Survey Statistics and Methodology, Iowa State University, Washington, DC/Ames

United States Environmental Protection Agency (2021) Inventory of U.S. greenhouse gas emissions and sinks: 1990–2019. EPA 430-R-21-005. U.S. Environmental Protection Agency, Washington, DC. https://www.epa.gov/ghgemissions/inventory-us-greenhouse-gas-emissions-and-sinks

Wentland SC, Ancona ZH, Bagstad KJ et al (2020) Accounting for land in the United States: integrating physical land cover, land use, and monetary valuation. Ecosyst Serv 46:101178. https://doi.org/10.1016/j.ecoser.2020.101178

Xu J, Morris PJ, Liu J, Holden J (2018) PEATMAP: refining estimates of global peatland distribution based on a meta-analysis. Catena 160:134–140. https://doi.org/10.1016/j.catena.2017.09.010

Yue C, Ciais P, Houghton RA, Nassikas AA (2020) Contribution of land use to the interannual variability of the land carbon cycle. Nat Commun 11:3170. https://doi.org/10.1038/s41467-020-16953-8

Zhu Z, McGuire AD (eds) (2016) Baseline and projected future carbon storage and greenhouse-gas fluxes in ecosystems of Alaska. U.S. geological survey professional paper 1826, 196 p. https://doi.org/10.3133/pp1826

Chapter 2
Soil Organic Carbon Stocks

Abstract The soil carbon (C) stock of terrestrial biomes is determined by natural (e.g., vegetation cover, soil type, climate) and anthropogenic (e.g., soil and land-use management) factors. The soil C stock is comprised of soil inorganic carbon (SIC) and soil organic carbon (SOC) stocks, with the latter mainly controlled by biome type. Data on SOC stocks for terrestrial biomes of the U.S. are often not directly available but rather for ecosystem-type groups, land use/land cover classes, land resource regions or land-use categories. The value of the SOC stock can be estimated based on the avoided social cost of C from the long-term damage resulting from carbon dioxide (CO_2) emissions as increases in the SOC stock contribute to climate change adaptation and mitigation. The SOC stock is also the most important master property determining the state of many soil physical, chemical and biological properties, and among the most important soil health indicators. However, anthropogenic conversions to managed land have resulted in SOC losses of 12.2 Pg C (13.4 billion tn C) in 0–200 cm (0–79 in) depth in the U.S. over the past 300 years. Adjustments in cropland management, in particular, can recover some of the lost SOC stock with an estimated total stock of 82.6 Pg C to 200-cm (79-in) depth for the conterminous U.S. (CONUS), i.e., the 48 adjoining U.S. states on the continent of North America. To this depth, forest biomes may contain up to 40 Pg SOC (44 billion tn SOC), but large uncertainty remains. Soils of the boreal forest/taiga biome in Alaska may contain SOC stocks >10 Pg C (11 billion tn C) in 0–1 m (0–7 ft) depth, and estimates need to be improved by better data for peatlands. The SOC stocks of the temperate coniferous forest, and the temperate broadleaf and mixed forest biomes are estimated at 12.2 Pg C (13.5 billion tn C) and 17.8 Pg C (19.7 billion tn C) in 0–150 cm (0–59 in) depth, respectively. To this soil depth, the SOC stock of the tropical forest biome of the U.S. may amount to 191 Tg C (211 million tn C). The SOC stocks of shrublands and rangelands are estimated at 5.6 Pg C (6.2 billion tn C) and 12.3 Pg C (13.6 billion tn C) in 0–100 cm (0–39 in) depth, respectively. Arctic and tundra SOC stocks in Alaska may amount to 19.2 and 21.6 Tg C (21.2 and 23.9 million tn C) to 100-cm (39-in) depth, respectively. Highly uncertain are the data for SOC stocks of terrestrial wetlands including peatlands. To 150-cm (59-in) depth, 28.5 Pg C (31.4 billion tn C) may be stored in wetlands and peatlands in CONUS and Alaska. Data for SOC stocks of the total area of deserts

K. Lorenz, R. Lal, *Soil Organic Carbon Sequestration in Terrestrial Biomes of the United States*, https://doi.org/10.1007/978-3-030-95193-1_2

and xeric shrublands in the U.S. have not been published. To 200-cm (7 ft) depth, cropland SOC stocks were estimated at 28.4 Pg C (31.3 billion tn C). Total SOC stock to 100-cm (39-in) depth for urban areas in CONUS was estimated to range between 1.67 and 1.84 Pg C (1.84–2.03 billion tn C). There is an urgent need to improve SOC stock estimates for the entire terrestrial biosphere of the U.S. to identify priority areas for climate change adaptation and mitigation, and enhancing soil quality and health by managing SOC sequestration and stocks. After a brief introduction, this chapter will summarize data on SOC stocks for the different biomes, ecosystems and land uses introduced in Chap. 1.

Keywords Soil carbon stock · Soil organic carbon stock · Monetary value · Soil properties · Soil health · Soil monitoring

2.1 Introduction

Production in terrestrial biomes, the main carbon (C) input to soil C, can be either limited by water or energy (Kraemer et al. 2020). Terrestrial biomes have soil C stocks determined by natural (e.g., vegetation cover, soil type, climate) and anthropogenic (e.g., soil and land-use management) factors. Principally, the soil C stock is comprised of soil inorganic carbon (SIC) and soil organic carbon (SOC) stocks (Lorenz and Lal 2018). Globally, biome is the main control of SOC stocks (De Deyn et al. 2008; Soudzilovskaia et al. 2019). Specifically, both edaphic and climatic factors may control the SOC stocks to 2-m (6.6 ft) depth globally (Luo et al. 2021). Losses of C from both the SIC and SOC stocks must be reduced and/or stocks increased by soil C sequestration as net increases in soil C stocks contribute to climate change adaptation and mitigation by storing atmospheric carbon dioxide (CO_2) in protected and stabilized fractions for millennia (Archer et al. 2009). Globally, mineral SOC has a mean age of 4830 years to 100-cm (39 in) depth inferred from radiocarbon measurements (Shi et al. 2020). In 0–30 cm (0–12 in) depth, mean SOC age ranges between 390 and 3490 years in tropical forest and tundra biomes, respectively. In 30–100 cm (12–39 in) depth, mean SOC age ranges between 2710 and 16,890 years in temperate forest and tundra biomes, respectively (Shi et al. 2020). The persistence of SOC and its responsiveness to soil and land use management is a global research priority. To enhance SOC persistence, Lehmann et al. (2020) proposed that soil management should focus on ongoing care to manipulate the intricate balance between C inputs and losses rather than rely on locking away C in soil for the long term. Both soil ecosystem functions that require decomposition and those that require stabilization by adsorption to mineral surfaces need maintenance (Hoffland et al. 2020). The potential tension between enhancing SOC sequestration *versus* improving soil properties need to be addressed (Waring et al. 2020).

In the introduction, this chapter presents the monetary value of SOC followed by discussions about the relationship of SOC with climate change, soil health and soil fertility. The following sections present discussions about historic losses of SOC and principal controls of the SOC stock. The main sections of this chapter summarize data on SOC stocks for the different biomes, ecosystems and land uses introduced in Chap. 1.

2.1.1 Monetary Value of the Soil Organic Carbon Stock

The SOC has monetary value for U.S. society (Mikhailova et al. 2019). The U.S. government's social cost of carbon (SCC) is defined as the monetary value of the damages caused by one additional metric ton of CO_2 emitted (Wagner 2021). Thus, SCC is a standard measure of the benefits of a climate policy (Aldy et al. 2021). Collectively, the social cost of greenhouse gases (GHGs) is the monetary value of the net harm to society associated with adding a small amount of that GHG to the atmosphere in a given year (IWG 2021). It includes the value of all climate change impacts, including changes in net agricultural productivity, human health effects, property damage from increased flood risk natural disasters, disruption of energy systems, risk of conflict, environmental migration, and the value of ecosystem services (IWG 2021).

A price per tonne of CO_2 emitted in 2020, in 2020 US dollars, of US$50 ($45 per ton of CO_2) has been proposed but criticized as too low (Wagner 2021). To limit global average warming to 1.5 °C (2.7 °F) at most, a price of US$100 per tonne (US$91 per ton) of CO_2 may be more consistent by 2030 (Wagner 2021). The value of SOC can particularly be assessed based on the avoided social cost of C from the long-term damage resulting from CO_2 emissions (Mikhailova et al. 2019). For the contiguous/conterminous U.S. (CONUS), i.e., the 48 adjoining U.S. states on the continent of North America, the avoided social cost of C has been estimated at $42 per tonne ($38 per ton) of CO_2 in 2007 U.S. dollars. This dollar figure represents the value of damages avoided due to an equivalent reduction or sequestration of CO_2. The total calculated monetary value of SOC storage in CONUS has been estimated to amount to $12.7 trillion U.S. dollars. The soil depth interval with the highest values of SOC storage and content was 20–100 cm (8–39 in) [$6.18 trillion and $0.84 m^{-2} ($9.0 ft^{-2}), respectively], while the depth interval 100–200 cm (39–79 in) had the lowest values of SOC storage ($2.88 trillion) and content [$0.39 m^{-2} ($4.2 ft^{-2})]. The depth trends exemplify the higher SOC stocks in the upper portions of the soil. The high monetary values contribute to the need for incentives and policies with regards to SOC management in CONUS (Mikhailova et al. 2019).

2.1.2 Soil Organic Carbon and Climate Change

The SOC stock has received much more attention in policy and practice than the SIC stock, and is, therefore, the focus of this book. The main natural C input to the SOC stock is the plant net primary production (NPP) entering the soil from growing roots (rhizodeposition), and deposition of shoot and root residues. Globally, belowground NPP is 46% of the total NPP, ranging from 32% for croplands to 64% for grasslands with other terrestrial biomes having intermediate values (Gherardi and Sala 2020). The addition of organic amendments such as biochar, biosolids, compost, manure and slurry can be additional C sources. The main natural SOC losses are caused by microbial decomposition of OM which releases CO_2, and also some methane (CH_4) back into the atmosphere (Lorenz and Lal 2018). Other losses of SOC result from charring or burning and volatilization and leaching of organic compounds (Amundson 2001).

Aside its importance for climate change adaptation and mitigation, the SOC content is the most important and universally accepted master property that determines the state of many soil physical (soil structure, density, porosity, water-holding capacity, percolation rate and erodibility), soil chemical (nutrient availability, sorption capacity and pH), and soil biological (biodiversity, microbial biomass and basal respiration) properties (Kuzyakov and Zamanian 2019). Thus, the importance of the SOC stock and its composition for soil functioning is an active field of research (Hoffland et al. 2020). SOC loss is among the major land degradation pathways (Prăvălie 2021).

2.1.3 Soil Organic Carbon and Soil Health

The SOC stock is also among the important soil health baseline indicators (Stott 2019). Human health depends on soil health (Brevik and Sauer 2015), because a healthy soil has "the capacity to function as a vital living system to sustain biological productivity, maintain environmental quality, and promote plant, animal, and human health" (Doran et al. 1996). However, soil health definitions often fail to fully capture the important role of soils as the largest reservoir of terrestrial C and essential nutrients for plant growth. Thus, for the agricultural community, Toor et al. (2021) defined soil health as the capacity of soils to provide a sink for C to mitigate climate change and a reservoir for storing essential nutrients for sustained ecosystem productivity. Soil health indicators refer to measurable attributes used to evaluate overall soil health or detect the effect of management practices on soil health. Farmers, however, do not need complex information to successfully grow crops with the two most essential needs of water and nutrient availability from the perspective of plants (Toor et al. 2021).

Healthy soil is required for nutritious food production; supporting infrastructure; participating in atmospheric exchanges of GHGs such as CO_2, nitrous oxide (N_2O), and CH_4; acting as water filter and purifier; and supporting a healthy micro- and macro-biome (Kemper and Lal 2017). Shelter, clothing and fuel are other ESs that a healthy soil provides to human health (Brevik et al. 2019). Most importantly in the era of accelerated climate change, healthy soil can sequester significant amounts of atmospheric CO_2 primarily through increases in the SOC stock. Thus, healthy soils with site-specific SOC stocks contribute to mitigating CO_2-related climate change (Lal 2016). To avoid dangerous climate impacts, SOC stocks that are not recoverable by direct, localized action on relevant timescales within 30 years must be protected (Goldstein et al. 2020). Boreal, temperate and tropical forests have no irrecoverable SOC in 0–30 cm (0–12 in) depth. Among grasslands, average irrecoverable SOC stock densities in 0–30 cm (0–12 in) depth are 0, 5 and 16 Mg C ha^{-1} (0, 2 and 7 tn C ac^{-1}) for tropical, temperate and montane grasslands, respectively. Highly vulnerable are peatlands with irrecoverable SOC stocks in 0–100 cm (0–39 in) depth of 135 Mg C ha^{-1} (60 tn C ac^{-1}) for boreal/temperate peatlands, and 450 Mg C ha^{-1} (201 tn C ac^{-1}) for tropical peatlands (Goldstein et al. 2020). In addition, a non-negligible proportion of the SIC stock is also irrecoverable (Zamanian et al. 2021).

2.1.4 Soil Organic Carbon and Soil Fertility

The SOC stock is also important for soil fertility as, for example, SOC is intimately associated with elements that are predominantly present in soil in organic forms, such as nitrogen (N) and phosphorus (P), as well as elements comprising the mineral matrix, such as silicon (Si), aluminum (Al), and iron (Fe; Nave et al. 2019). The connection of SOC with soil fertility spurred scientific interest in the U.S. for more than a century with the aim to manage and increase SOC stocks (Allison 1973). Many previous syntheses focused on the potential of U.S. soils to increase SOC with adoption of best management practices (BMPs) under different land uses (Chambers et al. 2016). The SOC sequestration rates were estimated at 0.3–0.5 Mg C ha^{-1} y^{-1} (megagram [0.1–0.2 US tn C ac^{-1} y^{-1}]), 0.04–0.21 Mg C ha^{-1} y^{-1} (0.02–0.09 tn C ac^{-1} y^{-1}), and 0.11–0.43 Mg C ha^{-1} y^{-1} (0.05–0.19 tn C ac^{-1} y^{-1}) for cropland, grazing land and forest land soils in the U.S., respectively (Lal et al. 1998; Follett et al. 2001; Kimble et al. 2003). This corresponded to total potentials of 45–98 Tg C y^{-1} (teragram [50–109 million tn C y^{-1}]), 13–70 Tg C y^{-1} (14–77 million tn C y^{-1}) and 25–102 Tg C y^{-1} (28–112 million tn C y^{-1}) for cropland, grazing land and forest land soils, respectively. Subsequently, more detailed assessments of the technical potential for SOC sequestration in U.S. agricultural top soils (0–30 cm [0–12 in] depth) were performed (Sperow et al. 2003; Sperow 2016, 2020). Total potential SOC increase from adoption of activities that increase SOC stocks (removing highly erodible land from crop production, including winter-cover crops, eliminating fallow, and no-till adoption) could reduce emissions into the atmosphere by an additional 47.3 Tg C y^{-1} (52.1 million tn C y^{-1}) for at least 20 years (Sperow 2020).

2.1.5 Historic Loss of Soil Organic Carbon

Historically, native terrestrial biomes and ecosystems in the U.S. had higher SOC stocks than the same regions store today. Losses occurred by anthropogenic conversion to managed land, particularly cropland and its management, primarily over the past 200–300 years as human population and land transformation increased substantially (National Academies of Sciences, Engineering, and Medicine 2019). For example, historically tilled soils in CONUS have lower total soil C and SOC stocks integrated to 30-cm (12-in) depth compared to soils without a history of tillage (Sulman et al. 2020). To this depth, ~210,000 km^2 (81,000 mi^2) of land in the U.S. recorded decreases in SOC in 2001–2015 (Prăvălie 2021).

The estimated total SOC losses for the U.S. are 3.0 Pg C (petagram [3.3 billion tn C]), 6.7 Pg C (7.4 billion tn C) and 12.2 Pg C (13.4 billion tn C) in 0–30 cm (0–12 in), 0–100 cm (0–39 in) and 0–200 cm (0–79 in) depth, respectively (Sanderman et al. 2017). These estimates of historical losses indicate a hypothetical though highly impractical upper bound for restoring SOC stocks by filling the gap between SOC capacity and current SOC stocks, and their particulate and mineral-associated fractions through adjustments to land management, specifically, in agro-ecosystems (Luo et al. 2020). However, the majority of soils of U.S. cropland and grassland have the capacity for increased SOC stocks consistent with the 4 per Thousand Initiative (4PT) which strives to address global climate change through the aspirational goal of enhancing the SOC stock on a large portion of the world's managed soils by an average annual increase of 0.4% (Le Foll 2015; Chambers et al. 2016). To realize this potential it would be important to focus on top soil (0–20 cm [0–8 in]) SOC stocks, and prioritize degraded lands, cropland, and grassland, because lower soil depths, and degraded lands, cropland, and grassland uses can be managed to increase SOC stocks and enhance SOC sequestration. However, not considered in the assessment of the capacity of U.S. soils is the potential of break-through genetic technologies like deep rooting crop phenotypes or application of off farm C sources like compost and biochar (Chambers et al. 2016). Further, the supply of nutrients such as N and P needed for SOC formation and sequestration, potential SOC stock saturation, and social and economic constraints must be addressed (Spohn 2020; Minasny et al. 2018).

Estimates for the SOC stock of the total land area of the U.S. and for CONUS are quite diverse (Table 2.1). Some estimates are based on the State Soil Geographic Database (STATSGO) for an area of 7.37 million km^2 (2.85 million mi^2) excluding urban land, bare rock, and other non-soil bodies (Guo et al. 2006a). Bliss et al. (2014) estimated SOC stocks based on the U.S. Soil Survey Geographic (SSURGO) database for an area of 6.69 million km^2 (2.58 million mi^2). Sanderman et al. (2017) used a machine learning-based model that was fitted using a global compilation of SOC data and the History Database of the Global Environment land use data in combination with climatic, landform and lithology covariates. Gonçalves et al. (2021) used geographically weighted regression based on georeferenced soil profile observations obtained from the Rapid Carbon Assessment. Global databases

Table 2.1 Historical losses of soil organic carbon for the total land area of the United States, and current stocks of the total land area and the conterminous United States (Gonçalves et al. 2021; Sanderman et al. 2017; Bliss et al. 2014; Guo et al. 2006a)

Soil depth		Total land area				CONUS	
		Loss		Stock		Stock	
cm	in	Pg C	Billion tn C	Pg C	Billion tn C	Pg C	Billion tn C
0–30	0–12	3.0	3.3	69.0	76.1	29.3	32.3
0–100	0–39	6.7	7.4	146.8	161.8	57.2–75.4	63.1–83.1
0–200	0–79	12.2	13.4	219.3	241.7	82.6	91.1
Total soil profile						73.4	80.9

Table 2.2 Depth distribution of soil organic carbon stocks for different terrestrial ecosystems in conterminous United States (Guo et al. 2006b)

Depth			Soil organic carbon stock	
cm	in	Terrestrial ecosystem	Pg C	Billion tn C
0–20	0–8	Agriculture	5.8	6.4
		Forest	7.3	8.1
		Grass	3.1	3.4
		Shrub	2.5	2.8
		Pasture	2.3	2.5
		Wetland	2.2	2.5
0–100	0–39	Agriculture	15.7	17.3
		Forest	18.2	20.0
		Grass	8.0	8.8
		Shrub	6.5	7.2
		Pasture	5.9	6.5
		Wetland	7.9	8.7
0–200	0–79	Agriculture	19.6	21.6
		Forest	23.1	25.5
		Grass	10.2	11.2
		Shrub	8.4	9.3
		Pasture	7.5	8.3
		Wetland	11.3	12.4

generally underestimate the SOC stock by 40% for SoilGrids and by 80–90% for the Harmonized World Soil Database (Tifafi et al. 2018).

Guo et al. (2006b) estimated SOC stocks in 0–20, 0–100 and 0–200 cm (0–8, 0–39 and 0–79 in) depths for soils under agriculture, forest, grass, pasture, shrub and wetland in the CONUS (Table 2.2). Among ecoregions, SOC stocks to 100-cm (39.4 in) depth ranged from 223.1 Tg C (245.9 million tn C) for Warm Deserts to 9557.8 Tg C (10.5 billion tn C) for Western Cordillera, respectively (Gonçalves et al. 2021).

2.1.6 Inventory of Soil Organic Carbon Stocks

There is no regular inventory of U.S. SOC stocks. Spencer et al. (2011) proposed the design of a national soil monitoring network for C on agricultural lands including determination of sample size, allocation, and site-scale plot design. Both, modeled estimates of SOC stock change variability and a set of soil sample measurements may serve as inputs to evaluate a potential network design for U.S. agricultural lands. Stratification by climate, soil, and land use with sites allocated based on modeled SOC stock change variability may effectively reduce the national standard error of SOC stock change (Spencer et al. 2011). The U.S. Rapid Carbon Assessment Project performed the first comprehensive one-time sampling of soils across CONUS (Wills et al. 2014). A multi-level hierarchical design was used to ensure that samples were distributed across regions, soils and land use/land cover classes. Within those strata, sites were selected at random locations where five pedons were described and sampled. Soil bulk density was calculated for samples from the upper 50 cm (20 in), and predicted for deeper samples using pedon and horizon information in a regression tree developed with random forests. The SOC concentration was predicted for each sample using processed Visible-Near Infrared spectra and a random forest model (Wills et al. 2014).

The U.S. Department of Agriculture Natural Resources Conservation Service (NRCS) compiled data for the U.S. Soil Survey Geographic (SSURGO) database which provides detailed soil mapping for most of CONUS. These data have been used to formulate estimates of soil C stocks from 1:24,000-scale or 1:12,000-scale maps. Some of the estimated SOC stocks are presented below separated for biomes, land cover and land-use classes (Bliss et al. 2014). These data can serve as baseline and reference to assess the spatial and temporal dynamics and sequestration of SOC stocks in CONUS in subsequent years. Other efforts to characterize SOC at national scale include the United States Department of Agriculture (USDA) Forest Service's Forest Inventory and Analysis (FIA) program, the National Cooperative Soil Survey (NCSS), and the International Soil Carbon Network (ISCN; Berryman et al. 2020). The NCSS program encompasses all lands of the U.S., including wetlands, forested areas, rangelands, and urban areas. Other key soils databases in the U.S. include State Soil Geographic (STATSGO2), and Long-term Soil Productivity (LTSP; Binkley et al. 2020). However, for Alaska only limited SSURGO data are available, and data for the STATSGO2 database are mapped at a coarser scale of 1:1,000,000 compared to a scale of 1:250,000 for the continental U.S., Hawaii, Puerto Rico, and the Virgin Islands (Kimsey et al. 2020). Conventional systems for mapping and classifying soils were not designed for estimating and managing stores and fluxes of SOC in the U.S., and a gridded product with three-dimensional (3D) estimates of soil properties is required instead (Ramcharan et al. 2018).

In the following section, data on SOC stocks for the U.S. will be presented based mainly on the biome classification and mapping by Olson et al. (2001). Olson et al. used expert knowledge and regional vegetation maps (in turn created by experts) to construct a global biome map (Conradi et al. 2020). However, some overlap cannot be avoided as there are no distinct boundaries between terrestrial biomes as they grade into each other with the intergradation called ecotone (Campbell and Reece 2005). For example, peat may occur within a forest, grassland, savanna, or wetland (Goldstein et al. 2020). Also, the SOC stock data are likely underestimations as sampling to 1-m (39-in) depth should be but is often not standard, and even deeper sampling may be necessary in certain cases (Gross and Harrison 2019). Further, SOC data are not available for some of the biomes but rather for ecosystem-type groups, land use/land cover classes, land resource regions or land-use categories. Another issue are differences between the cited studies in accuracy of SOC stock assessments for generally not flat land surfaces, i.e., >60% of Earth's land surface is composed of landscapes with >8% slope (Staub and Rosenzweig 1992). Particularly in mountain soils, various properties with relevance for SOC stocks have a large spatial variability, including soil depth, stoniness, bulk density, and SOC concentration (Prietzel and Wiesmeier 2019). Thus, it is assumed that the reported large-scale SOC stock assessments are based on sampling points representative for the studied landscape and adequate techniques are applied for regionalization of SOC stock data of soil profiles (Prietzel and Wiesmeier 2019).

2.1.7 Principal Controls of Soil Organic Carbon

Temperature, precipitation and geochemical variables may be suitable for predicting SOC concentrations and trends with soil depth in the U.S. (Yu et al. 2021). Specifically, climate was a fundamental predictor of SOC concentration to 1-m (39-in) depth, and played similarly important roles as some geochemical predictors based on data from 2574 mineral horizons from 675 pits from the National Ecological Observatory Network sites. Soil depth, oxalate-extractable Al (Al_{ox}), pH, and exchangeable calcium plus exchangeable magnesium were important while silt + clay, oxalate-extractable Fe (Fe_{ox}), and vegetation type were weaker predictors. The relative importance of geochemical and climate predictors for SOC concentration with increasing soil depth was mostly constant. Together, climate and geochemical factors are jointly important in mediating spatial distribution of SOC concentration over continental scales in the U.S. (Yu et al. 2021).

The largest control on the SOC stocks distribution in 0–30 cm (0–12 in) depth in CONUS is governed by temperature, followed by land use and land cover, and topography (Adhikari et al. 2020). In addition to these environmental predictors, drainage, soil type, precipitation, ecological zone, and surficial geology, all representing the major soil-forming factors are significant predictors of topsoil SOC stocks at different spatial scales. The SOC stock distribution to 30-cm (12-in) depth in CONUS is highly variable (CV = 157%), with a mean and standard deviation of 95 and 150 Mg C ha^{-1} (42 and 67 tn C ac^{-1}), respectively (Adhikari et al. 2020). Higher precipitation and lower temperatures were associated with higher levels of SOC stocks to 100-cm (39 in) depth in the majority of CONUS ecoregions (Gonçalves et al. 2021). Changes in land cover types (vegetation properties) was important in drier ecosystems such as deserts, whereas soil types and topography were more important in the prairies (Gonçalves et al. 2021).

2.2 Forest Biomes

About 20 Pg C (22 billion tn C) are stored in all live trees (aboveground and belowground) in CONUS (Domke et al. 2020). In comparison, the total SOC stock in 2012 was estimated at 15 Pg C (17 billion tn C), and it potentially remained almost stable since 1907 (Magerl et al. 2019). This amount corresponds well to the estimates by Woodall et al. (2015) and Heath et al. (2011) while Sleeter et al. (2018) and Liu et al. (2020b) reported higher SOC stocks of 27 Pg C (30 billion tn C) and 40 Pg C (44 billion tn C), respectively, including SOC stocks to a depth of 2 m (7 ft).

The forest land SOC storage in the U.S. is monitored by the National Forest Inventory (NFI) conducted by the FIA program within the USDA Forest Service (O'Neill et al. 2005). The FIA program provides official estimates of forest C stocks and flows for the U.S., which are reported annually in the U.S. Greenhouse Gas Inventory under the United Nations Framework Convention on Climate Change (UNFCCC; Janowiak et al. 2017). The FIA program has consistently measured soil attributes as part of the NFI since 2001, and has compiled an extensive inventory of SOC in forest land in CONUS, and southeast and southcentral coastal Alaska. In 2019, mineral soil SOC stocks increased by 0.7 Tg C (772,000 tn C) while organic soils lost 0.3 Tg C (331,000 tn C; Land-use category 'Forest Land Remaining Forest Land'; U.S. EPA 2021). To 1-m (3-ft) depth, an estimated 25.1 Pg C (27.7 billion tn C) were stored as SOC in mineral soil, and 5.9 Pg C (6.5 billion tn C) in organic soil in CONUS and Alaska in 2019. However, the SOC stock of forests in the U.S. is highly variable and much uncertainty remains (Domke et al. 2017).

Unclear is the response of U.S. forests to increasing atmospheric CO_2 concentrations and to increases in temperature. For example, NPP could go up or stay unchanged, and tree biomass could go up or down depending on projected increases in temperature and atmospheric CO_2 concentrations. Young aggrading forests showed a strong CO_2 fertilization effect on biomass growth, but the additional C uptake of mature forests at elevated CO_2 levels may not lead to increased C sequestration at the ecosystem level (Jiang et al. 2020). For example, the majority of the extra C tracked in the first ecosystem-scale Free-Air CO_2 Enrichment (FACE) experiment in a mature forest was emitted back into the atmosphere via respiratory fluxes, with increased soil respiration alone accounting for half of the total uptake surplus (Jiang et al. 2020). For boreal forests, a lower CO_2 fertilization and biomass increase is projected in the future (Sperry et al. 2019). Otherwise, some foliar nutrient concentrations in northern European forests increased during the last three decades with favorable conditions of mean annual precipitation and temperature (Penuelas et al. 2020). Thus, the C capture of northern forests in the U.S. may also potentially improve in the future due to improved nutritional status. However, the nutritional status of Mediterranean and temperate forests in Europe declined due to increasing atmospheric CO_2 concentrations (Penuelas et al. 2020). This declining nutritional status may have negative feedbacks on Mediterranean and temperate forest C capture also in the US.

In the following sections, SOC stocks for U.S. forests are reported for forest-type groups, land use/land cover classes, land resource regions or land-use categories as those are generally not reported for forest biomes.

2.2.1 Boreal Forest/Taiga Biome

Boreal forests contain a large, slow-cycling SOC stock which exceeds the vegetation C stock (Amundson 2001). Peat deposits, in particular, store more C than trees in forested peatlands of the boreal biome (Beaulne et al. 2021). However, data on SOC stocks for the total boreal forest area of the U.S. are not available (Table 2.3). Assuming that SOC densities in the boreal forest area are similar to the 55.8 Mg C ha^{-1} (24.9 tn C ac^{-1}) in litter, live moss, organic layer and 2.54 cm (1 in) depth of mineral soil of a pilot inventory covering 0.010 million km^2 (0.004 million mi^2) in interior Alaska (Pattison et al. 2018), it can be estimated that boreal forest surface soils of interior Alaska contain >2.5 Pg C (2.7 billion tn C). However, it is likely that substantial amounts of mineral SOC are present at greater soil depths (Johnson and Kern 2003). For example, Douglas et al. (2014) summarized SOC stocks for representative interior Alaska ecosystem types ranging from 72 Mg C ha^{-1} (32 tn C ac^{-1}) for deciduous forests to 513 Mg C ha^{-1} (229 tn C ac^{-1}) for black spruce (*Picea mariana* Mill BSP) lowland ecosystems. Thus, organic layers and mineral soils to 1-m (7-ft) depth in the boreal forest biome in interior Alaska may contain between 5.0 Pg C (5.4 billion tn C) and 7.5 Pg C (8.1 billion tn C; Pattison et al. 2018). In comparison, Birdsey (1992) estimated that forest soils in Alaska contain 10.1 Pg C (11.1 billion tn C) to 1-m (7-ft) depth. Most boreal C in Alaska resides in peatlands and soils but estimates are highly uncertain (Bradshaw and Warkentin 2015). There are only limited soil data available for Alaska (Kimsey et al. 2020).

Table 2.3 Soil organic carbon stocks of areas dominated by boreal forest land-cover types, boreal forests of Alaska, and the total forest area of Alaska (Zhu and McGuire 2016; Johnson and Kern 2003)

	Soil depth		Soil organic carbon stock	
Region	cm	in	Pg C	Billion tn C
Boreal forest land-cover types	Not specified	Not specified	18.9	20.8
Boreal forests of Alaska	Not specified	Not specified	10.0–30.8	11.0–34.0
Total forest area Alaska	0–30	0–12	4.4	4.9
	0–100	0–39	10.4	11.5
	0–150	0–59	13.8	15.2

2.2.2 Temperate Coniferous Forest Biome

Johnson and Kern (2003) determined the spatial patterns and total amounts of SOC in temperate forests of the U.S. using moderately detailed digital soil-survey data (National Soil Characterization Database NSCD) linked to a large national soil-characterization database (State Soil Geographic STATSGO soil mapping) using soil classification. The SOC stocks were estimated at 6.3 Pg C (6.9 billion tn C), 11.5 Pg C (12.7 billion tn C) and 12.2 Pg C (13.5 billion tn C) in 0–30 cm (0–12 in), 0–100 cm (0–39 in) and 0–150 cm (0–59 in) depth, respectively (Johnson and Kern 2003). In contrast to previous assessments, those estimates included Histosols, considered the full soil profile, and made corrections for coarse fragments and bulk density for the entire U.S. (Johnson and Kern 2003).

2.2.3 Temperate Broadleaf and Mixed Forest Biome

The SOC stocks for the temperate broadleaf and mixed forest biome of the U.S. were also estimated by Johnson and Kern (2003). Estimates ranged between 7.2 Pg C (8.0 billion tn C) in 0–30 cm (0–12 in) depth, 13.4 Pg C (14.8 billion tn C) in 0–100 cm (0–39 in) depth, and 17.8 Pg C (19.7 billion tn C) in 0–150 cm (0–59 in) depth, respectively (Johnson and Kern 2003).

2.2.4 Tropical Forest Biome

The tropical forests of Hawaii (mixed and native) contain estimated SOC stocks of 44.9 Tg C (49.5 million tn C), 96.1 Tg C (105.9 million tn C) and 117.9 Tg C (130.0 million tn C) in 0–30 cm (0–12 in), 0–100 cm (0–39 in) and 0–150 cm (0–59 in) depth, respectively (Johnson and Kern 2003). In comparison, Selmants et al. (2017) reported SOC stocks of 97.0 Tg C (106.9 million tn C) in 0–100 cm (0–39 in) depth for native dry forest, nonnative dry forest, native mesic-wet forest, invaded mesic-wet forest, and plantations with nonnative tree species in Hawaii.

The estimated SOC stocks of forests in Puerto Rico were 22.8 Tg C (25.1 million tn C), 32.8 Tg C (36.2 million tn C) and 37.0 Tg C (40.8 million tn C) in 0–30 cm (0–12 in), 0–100 cm (0–39 in) and 0–150 cm (0–59 in) depth, respectively (Johnson and Kern 2003). Assuming that SOC stock densities are similar to the those in Puerto Rico, it can be estimated that SOC stocks in the U.S. Virgin Islands amount to 0.9 Tg C (1.0 million tn C), 1.3 Tg C (1.4 million tn C) and 1.4 Tg C (1.5 million tn C) in 0–30 cm (0–12 in), 0–100 cm (0–39 in) and 0–150 cm (0–59 in) depth, respectively (Johnson and Kern 2003; Oswalt et al. 2019). Under the assumption that SOC stock densities are similar to those in Hawaii, it can further be estimated that SOC stocks of the U.S. Pacific Islands except Hawaii amount to 13.4 Tg C

(14.7 million tn C), 28.6 Tg C (31.5 million tn C) and 35.1 Tg C (38.7 million tn C) in 0–30 cm (0–12 in), 0–100 cm (0–39 in) and 0–150 cm (0–59 in) depth, respectively (Johnson and Kern 2003; Oswalt et al. 2019).

2.3 Grassland and Shrubland Biomes

2.3.1 *Temperate Grasslands, Savannas, and Shrublands Biome*

Up to 80% of total grassland ecosystem C may be found in soil (Janowiak et al. 2017). Temperate grassland SOC stocks are highest in regions were rainfall is the greatest such as the tallgrass prairie in the humid temperate region of the U.S. Otherwise, temperate grassland SOC stocks decrease with increasing annual temperature due to greater evapotranspiration (Janowiak et al. 2017). Semiarid grasslands are C neutral while semiarid savannas and shrublands are C sinks (Zhang et al. 2020). The SOC stocks for shrublands (shrub/scrub) in the U.S. amount to an estimated 3.0 Pg C (3.4 billion tn C) and 5.6 Pg C (6.2 billion tn C) in 0–30 cm (0–12 in) and 0–100 cm (0–39 in) soil depth, respectively (Bliss et al. 2014). Rangelands (grass/pasture/hay) are estimated to contain 6.3 Pg C (7.0 billion tn C) and 12.3 Pg C (13.6 billion tn C) in 0–30 cm (0–12 in) and 0–100 cm (0–39 in) soil depth, respectively (Bliss et al. 2014). Sundquist et al. (2009) estimated SOC stocks of 9.7 Pg C (10.7 billion tn C) and 11.2 Pg C (12.3 billion tn C) to 100-cm (39 in) depth for shrublands, and grasslands and other herbaceous areas, non-vegetated areas, and open water, respectively.

2.3.2 *Tundra Biome*

A unique feature of C stocks in tundra is the predominance of SOC as a proportion of the total tundra C stock as biomass production is limited by harsh environmental conditions and short growing seasons (Schuur et al. 2018). Thus, soil C input in tundra is slow with a high proportion of plant C allocated belowground but mainly in the soil surface due to permafrost (De Deyn et al. 2008). Root-to-total tundra biomass ratios are 0.62 compared to 0.27 for those of boreal forests (Saugier et al. 2001). Soil C output is also slow due to poor litter quality, low soil temperatures and often soil anoxia, resulting in old and large C stocks in poorly decomposed litter (De Deyn et al. 2008). Tundra soils act as long-term (i.e., century to millennia) C sinks as C continues to accumulate as dead OM (Schuur et al. 2018). The measured and modelled arctic and alpine tundra SOC stocks in Alaska are 19.2 and 21.6 Tg C (21.2 and 23.9 million tn C) to 1-m (39-in) soil depth based on data for area and SOC stocks of tundra land-cover types reported by Zhu and McGuire (2016). However, arctic tundra SOC stock data are uncertain as deeper-soil profiles and cryoturbation are often not considered (Ping et al. 2008). Further, data on alpine tundra SOC stocks for CONUS are not available.

2.4 Wetland Biomes

2.4.1 *Terrestrial Wetlands*

The data for SOC stocks of terrestrial wetlands including peatlands are uncertain (Table 2.4). To 150-cm (59-in) depth, 28.5 Pg C (31.4 billion tn C) may be stored in total in wetlands and peatlands in CONUS and Alaska. Aside uncertainty in acreage, the total depth of peat is seldomly sampled with typical assessment depths of 100–200 cm (39–79 in; Kolka et al. 2018). Based on extensive peat core data, mean northern peatland depth was estimated at 249 cm (98 in) with permafrost-free peatlands deeper (286 cm [113 in]) than permafrost-affected peatlands (205 cm [81 in]; Hugelius et al. 2020). About 46% of northern peatlands are underlain by permafrost. It is likely that the SOC stock data reported in Table 2.3 increase if the entire peat depth is considered. On average, the net C flux of U.S. wetlands is negative indicating that they are currently C sinks (Kolka et al. 2018). However, wetlands including both inland freshwater wetlands and coastal wetlands dominated the soil CH_4

Table 2.4 Soil organic carbon stock of terrestrial wetlands and peatlands in different regions of the United States

Type	Region	Soil depth		Soil organic carbon stock		References
		cm	in	Pg C	Billion tn C	
Woody and emergent herbaceous wetlands	Conterminous U.S.	0–30	0–12	3.7	4.1	Bliss et al. (2014)
		0–100	0–39	8.9	9.8	
Freshwater inland wetlands	Conterminous U.S.	0–100	0–39	10.7	11.8	Nahlik and Fennessy (2016)
Mineral soil wetlands	Conterminous U.S.	0–150	0–59	4.8[a]	5.3[a]	Kolka et al. (2018)
	Alaska	0–150	0–59	10.5[a]	11.6[a]	Bridgham et al. (2006)
		0–200	0–79	26.0	28.7	
Peatlands	Conterminous U.S.	0–150	0–59	7.7[a]	8.5[a]	Kolka et al. (2018), Bridgham et al. (2006) and Joosten (2010)
		0–200	0–79	14.0	15.4	
		0–200	0–79	13.7	15.1	
	Alaska	0–150	0–59	5.5[a]	6.0[a]	Kolka et al. (2018), Joosten (2010) and Bridgham et al. (2006)
		0–200	0–79	15.5	17.1	

[a]Assuming that soil organic carbon accounts for 93% of the total carbon stock reported in Kolka et al. (2018)

emissions in CONUS in the 2000s (15.3 Tg CH_4 y^{-1} [16.9 million tn CH_4 y^{-1}]; Shu et al. 2020). Overall, peatlands exert a biogeochemical cooling effect through long-term sequestration of CO_2 while boreal peatlands have the potential to mitigate the effect of regional climate warming during the growing season through biophysical climate impacts (Helbig et al. 2020).

2.5 Desert and Xeric Shrubland Biome

Data for the SOC stocks of the total area of deserts and xeric shrublands in the U.S. have not been published. In comparison, the SOC stocks for barren land cover in CONUS amounted to an estimated 97 Tg C (107 million tn C) and 197 Tg C (217 million tn C) in 0–30 cm (0–12 in) and 0–100 cm (0–39 in) soil depth, respectively (Bliss et al. 2014). The shrub/scrub land cover SOC stocks were estimated at 3.0 Pg C (3.4 billion tn C) and 5.6 Pg C (6.2 billion tn C) in 0–30 cm (0–12 in) and 0–100 cm (0–39 in) soil depth, respectively (Bliss et al. 2014).

2.6 Cropland

The crop land cover SOC stocks in CONUS amounted to an estimated 6.8 Pg C (7.5 billion tn C) and 13.4 Pg C (14.7 billion tn C) in 0–30 cm (0–12 in) and 0–100 cm (0–39 in) soil depth, respectively (Bliss et al. 2014). In comparison, USDA's Rapid Carbon Assessment reported a stock of 13.0 Pg SOC (14.3 billion tn SOC) to 100-cm (39-in) depth for the land-use class Agriculture (Lajtha et al. 2018). To 200-cm (7 ft) depth, cropland SOC stocks were estimated at 28.4 Pg C (31.3 billion tn C) for the CONUS (Liu et al. 2020b). For U.S. cropland, Zomer et al. (2017) estimated a stock of 18.9 Pg SOC (20.8 billion tn C) to 30-cm (12-in) depth for available cropland soils (i.e., those not excluded as high SOC or sandy soils).

2.7 Urban Areas

Satellite-based approaches are essential for mapping changes in urban extent, but higher-resolution satellite-derived urban land-cover maps have become available only recently (Liu et al. 2020a). Thus, urban land area data estimations for the U.S. are insufficient for C calculations which is among the reasons for poor quantification of urban SOC stocks (Bliss et al. 2014; U.S. EPA 2021). For example, estimates of total SOC stocks to 1-m (3 ft) depth for the land-use class 'Urban' in CONUS range between 1.9 and 3.3 Pg C (2.1 and 3.6 billion tn C; Lajtha et al. 2018). The low estimate was based on an average density of 7.7 kg SOC m^{-2} (1.6

lbs SOC ft^{-2}) and a total urban land area of 252,000 km^2 (96,526 mi^2; Pouyat et al. 2006). In contrast, the high estimate was based on an average density of 8.5 kg SOC m^{-2} (1.7 lbs SOC ft^{-2}) and a total area of developed land of 383,375 km^2 (148,022 mi^2; Bliss et al. 2014). For 0–30 cm (0–12 in) depth, the estimated average density for developed land was 4.5 kg SOC m^{-2} (0.9 lbs ft^{-2}) with a total storage of 1.7 Pg SOC (1.9 billion tn SOC; Bliss et al. 2014). Churkina et al. (2010) estimated SOC stocks to 1-m (3 ft) depth for human settlements in CONUS of 0.4 Pg C (0.4 billion tn C) on 95,018 km^2 (36,687 mi^2) urban area, and of 9.4 Pg C (10.4 billion tn C) on 1,395,347 km^2 (538,747 mi^2) of exurban area. The area of human settlements (urban plus exurban) was defined using the density of housing units in census block-groups (Churkina et al. 2010).

The difference in population density between urban areas and dense settlements in the U.S. may not affect estimates for SOC stocks as was also reported for urban areas globally (Scharenbroch et al. 2018). Thus, SOC stocks to 1-m (3 ft) depth of urban areas for the U.S. may also be estimated based on anthropogenic biome areas (Ellis and Ramankutty 2008), and on SOC densities reported in Pouyat et al. (2006) and Bliss et al. (2014). Total SOC stock to 1-m (3-ft) depth for the CONUS urban area was estimated at 1.67–1.84 Pg C (1.84–2.03 billion tn C). The anthropogenic biome Dense Settlements was estimated to contain 1.91–2.11 Pg C (2.11–2.33 billion tn C) as SOC to 1-m (3-ft) depth.

2.8 Conclusions

Terrestrial biomes of the U.S. contain precious SOC stocks important for climate change adaptation and mitigation, and for maintaining and improving soil properties and soil health. However, the total SOC stock may be higher today if land clearing and soil disturbance for agriculture would have not occurred in the past. Some of the SOC lost as CO_2 into the atmosphere can potentially be recaptured by soil and land-use management practices resulting in increases in the SOC stock. However, peatlands in boreal and temperate regions have large irrecoverable SOC stocks. The largest SOC stock in the U.S. is found in forest biomes followed by wetland/peatlands, grasslands and croplands. However, data on biome SOC stocks are uncertain given the high spatial variability of SOC, the extensive land area of the U.S., the unclear delineation of terrestrial biome boundaries, and missing empirical assessments of entire soil profiles.

References

Adhikari K, Mishra U, Owens PR et al (2020) Importance and strength of environmental controllers of soil organic carbon changes with scale. Geoderma 375:114472. https://doi.org/10.1016/j.geoderma.2020.114472

Aldy JE, Kotchen MJ, Stavins RN, Stock JH (2021) Keep climate policy focused on the social cost of carbon. Science 373:850–852. https://doi.org/10.1126/science.abi7813

Allison FE (1973) Soil organic matter and its role in crop production, Developments in soil science 3. Elsevier Scientific, Amsterdam

Amundson R (2001) The carbon budget in soils. Annu Rev Earth Planet Sci 29:535–562

Archer D, Eby M, Brovkin V et al (2009) Atmospheric lifetime of fossil fuel carbon dioxide. Annu Rev Earth Planet Sci 37:117–134. https://doi.org/10.1146/annurev.earth.031208.100206

Beaulne J, Garneau M, Magnan G, Boucher É (2021) Peat deposits store more carbon than trees in forested peatlands of the boreal biome. Sci Rep 11:2657. https://doi.org/10.1038/s41598-021-82004-x

Berryman E, Hatten J, Page-Dumroese DS et al (2020) Soil carbon. In: Pouyat R, Page-Dumroese D, Patel-Weynand T, Geiser L (eds) Forest and rangeland soils of the United States under changing conditions. Springer, Cham, pp 9–31. https://doi.org/10.1007/978-3-030-45216-2_2

Binkley D, Richter DD, Pouyat RV, Geiser LH (2020) State of forest and rangeland soils research in the United States. In: Pouyat R, Page-Dumroese D, Patel-Weynand T, Geiser L (eds) Forest and rangeland soils of the United States under changing conditions. Springer, Cham, pp 1–7. https://doi.org/10.1007/978-3-030-45216-2_1

Birdsey RA (1992) Carbon storage and accumulation in US forest ecosystems. General Technical Report W0-59, USDA Forest Service, Washington, DC, 51 p

Bliss NB, Waltman SW, West LT, Neale A, Mehaffey M (2014) Distribution of soil organic carbon in the conterminous United States. In: Hartemink A, McSweeney K (eds) Soil carbon, Progress in soil science. Springer, Cham, pp 85–93. https://doi.org/10.1007/978-3-319-04084-4_9

Bradshaw CJA, Warkentin IG (2015) Global estimates of boreal forest carbon stocks and flux. Glob Planet Chang 128:24–30. https://doi.org/10.1016/j.gloplacha.2015.02.004

Brevik EC, Sauer TJ (2015) The past, present, and future of soils and human health studies. Soil 1:35–46. https://doi.org/10.5194/soil-1-35-2015

Brevik EC, Pereg L, Pereira P et al (2019) Shelter, clothing, and fuel: often overlooked links between soils, ecosystem services, and human health. Sci Total Environ 651:134–142. https://doi.org/10.1016/j.scitotenv.2018.09.158

Bridgham SD, Megonigal JP, Keller JK, Bliss NB, Trettin C (2006) The carbon balance of North American wetlands. Wetlands 26:889–916

Campbell NA, Reece JB (2005) Biology, 7th edn. Pearson, Benjamin Cummings, San Francisco

Chambers A, Lal R, Paustian K (2016) Soil carbon sequestration potential of US croplands and grasslands: implementing the 4 per thousand initiative. J Soil Water Conserv 71:68A–74A. https://doi.org/10.2489/jswc.71.3.68A

Churkina G, Brown DG, Keoleian G (2010) Carbon stored in human settlements: the conterminous United States. Glob Chang Biol 16:135–143. https://doi.org/10.1111/j.1365-2486.2009.02002.x

Conradi T, Slingsby JA, Midgley GF et al (2020) An operational definition of the biome for global change research. New Phytol 227:1294–1306. https://doi.org/10.1111/nph.16580

De Deyn CB, Cornelissen JHC, Bardgett RD (2008) Plant functional traits and soil carbon sequestration in contrasting biomes. Ecol Lett 11:516–531. https://doi.org/10.1111/j.1461-0248.2008.01164.x

Domke GM, Perry CH, Walters BF et al (2017) Toward inventory-based estimates of soil organic carbon in forests of the United States. Ecol Appl 27:1223–1235. https://doi.org/10.1002/eap.1516

Domke GM, Oswalt SN, Walters BF, Morin RS (2020) Tree planting has the potential to increase carbon sequestration capacity of forests in the United States. Proc Natl Acad Sci U S A 117:24649–24651. https://doi.org/10.1073/pnas.2010840117

Doran JW, Sarrantonio M, Liebig MA (1996) Soil health and sustainability. Adv Agron 56:1–54

Douglas TA, Jones MC, Hiemstra CA, Arnold JR (2014) Sources and sinks of carbon in boreal ecosystems of interior Alaska: a review. Elementa Sci Anthropocene 2:00032. https://doi.org/10.12952/journal.elementa.000032

Ellis EC, Ramankutty N (2008) Putting people in the map: anthropogenic biomes of the world. Front Ecol Environ 6:439–447. https://doi.org/10.1890/070062

Follett RF, Kimble JM, Lal R (2001) The potential of U.S. grazing lands to sequester soil carbon. In: Follett R, Kimble JM, Lal R (eds) The potential of US grazing lands to sequester carbon and mitigate the greenhouse effect. CRC/Lewis Publishers, Boca Raton, pp 401–430

Gherardi LA, Sala OE (2020) Global patterns and climatic controls of belowground net carbon fixation. Proc Natl Acad Sci U S A 117:20038–20043. www.pnas.org/cgi/doi/10.1073/pnas.2006715117

Goldstein A, Turner WR, Spawn SA et al (2020) Protecting irrecoverable carbon in Earth's ecosystems. Nat Clim Chang 10:287–295. https://doi.org/10.1038/s41558-020-0738-8

Gonçalves DRP, Mishra U, Wills S, Gautam S (2021) Regional environmental controllers influence continental scale soil carbon stocks and future carbon dynamics. Sci Rep 11:6474. https://doi.org/10.1038/s41598-021-85992-y

Gross CD, Harrison RB (2019) The case for digging deeper: soil organic carbon storage, dynamics, and controls in our changing world. Soil Syst 3:28. https://doi.org/10.3390/soilsystems3020028

Guo Y, Amundson R, Gong P, Yu Q (2006a) Quantity and spatial variability of soil carbon in the conterminous United States. Soil Sci Soc Am J 70:590–600. https://doi.org/10.2136/sssaj2005.0162

Guo Y, Gong P, Amundson R, Yu Q (2006b) Analysis of factors controlling soil carbon in the conterminous United States. Soil Sci Soc Am J 70:601–612. https://doi.org/10.2136/sssaj2005.0163

Heath LS, Smith JE, Skog KE, Nowak DJ, Woodall CW (2011) Managed forest carbon estimates for the US greenhouse gas inventory, 1990–2008. J For 109:167–173

Helbig M, Waddington JM, Alekseychik P et al (2020) The biophysical climate mitigation potential of boreal peatlands during the growing season. Environ Res Lett 15:104004. https://doi.org/10.1088/1748-9326/abab34

Hoffland E, Kuyper TW, Comans RNJ, Creamer RE (2020) Eco-functionality of organic matter in soils. Plant Soil 455:1–22. https://doi.org/10.1007/s11104-020-04651-9

Hugelius G, Loisel J, Chadburn S et al (2020) Large stocks of peatland carbon and nitrogen are vulnerable to permafrost thaw. Proc Natl Acad Sci U S A 117:20438–20446. www.pnas.org/cgi/doi/10.1073/pnas.1916387117

Interagency Working Group (IWG) on Social Cost of Greenhouse Gases, United States Government (2021) Technical support document: social cost of carbon, methane, and nitrous oxide. Interim estimates under executive order 13990. https://www.whitehouse.gov/wp-content/uploads/2021/02/TechnicalSupportDocument_SocialCostofCarbonMethaneNitrousOxide.pdf?source=email

Janowiak M, Connelly WJ, Dante-Wood K et al (2017) Considering forest and grassland carbon in land management. USDA Forest Service General Technical Report WO-95

Jiang M, Medlyn BE, Drake JE et al (2020) The fate of carbon in a mature forest under carbon dioxide enrichment. Nature 580:227–231. https://doi.org/10.1038/s41586-020-2128-9

Johnson MG, Kern JS (2003) Quantifying the organic carbon held in forested soils of the United States and Puerto Rico. In: Kimble JS, Lal R, Birdsey R, Heath LS (eds) The potential of U.S. forest soils to sequester carbon and mitigate the greenhouse effect. CRC Press LLC, Boca Raton, p 47–72

Joosten H (2010) The global peatland CO_2 picture. Peatland status and drainage related emissions in all countries of the world. Wetlands International. https://unfccc.int/files/kyoto_protocol/application/pdf/draftpeatlandco2report.pdf

Kemper KJ, Lal R (2017) Pay dirt! Human health depends on soil health. Complement Ther Med 32:A1–A2. https://doi.org/10.1016/j.ctim.2017.04.005

Kimble J, Birdsey R, Heath L, Lal R (eds) (2003) The potential of US forest soils to sequester carbon and mitigate the greenhouse effect. CRC Press, Boca Raton

Kimsey MJ, Laing LE, Anderson SM et al (2020) Soil mapping, monitoring, and assessment. In: Pouyat R, Page-Dumroese D, Patel-Weynand T, Geiser L (eds) Forest and rangeland soils of the United States under changing conditions. Springer, Cham, pp 169–188. https://doi.org/10.1007/978-3-030-45216-2_9

Kolka R, Trettin C, Tang W et al (2018) Chapter 13: terrestrial wetlands. In: Cavallaro N, Shrestha G, Birdsey R et al (eds) Second state of the carbon cycle report (SOCCR2): a sustained assessment report. U.S. Global Change Research Program, Washington, DC, pp 507–567. https://doi.org/10.7930/SOCCR2.2018.Ch13

Kraemer G, Camps-Valls G, Reichstein M, Mahecha MD (2020) Summarizing the state of the terrestrial biosphere in few dimensions. Biogeosciences 17:2397–2424. https://doi.org/10.5194/bg-17-2397-2020

Kuzyakov Y, Zamanian K (2019) Reviews and syntheses: Agropedogenesis – humankind as the sixth soil-forming factor and attractors of agricultural soil degradation. Biogeosciences 16:4783–4803. https://doi.org/10.5194/bg-16-4783-2019

Lajtha K, Bailey VL, McFarlane K et al (2018) Chapter 12: soils. In: Cavallaro N, Shrestha G, Birdsey R et al (eds) Second state of the carbon cycle report (SOCCR2): a sustained assessment report. U.S. Global Change Research Program, Washington, DC, pp 469–506. https://doi.org/10.7930/SOCCR2.2018.Ch12

Lal R (2016) Soil health and carbon management. Food Energy Secur 5:212–222. https://doi.org/10.1002/fes3.96

Lal R, Kimble JM, Follett R, Cole CV (1998) The potential of US cropland to sequester carbon and mitigate the greenhouse effect. Sleeping Bear Press, Chelsea

Le Foll S (2015) 4 per 1000: a new program for carbon sequestration in agriculture. French Minister of Agriculture, Agrifood and Forestry, Paris

Lehmann J, Hansel CM, Kaiser C et al (2020) Persistence of soil organic carbon caused by functional complexity. Nat Geosci 13:529–534. https://doi.org/10.1038/s41561-020-0612-3

Liu X, Huang Y, Xu X et al (2020a) High-spatiotemporal-resolution mapping of global urban change from 1985 to 2015. Nat Sustain 3:564–570. https://doi.org/10.1038/s41893-020-0521-x

Liu J, Sleeter BM, Zhu Z et al (2020b) Critical land change information enhances the understanding of carbon balance in the United States. Glob Chang Biol 26:3920–3929. https://doi.org/10.1111/gcb.15079

Lorenz K, Lal R (2018) Carbon sequestration in agricultural ecosystems. Springer, Cham

Luo Z, Viscarra-Rossel RA, Shi Z (2020) Distinct controls over the temporal dynamics of soil carbon fractions after land use change. Glob Chang Biol 26:4614–4625. https://doi.org/10.1111/gcb.15157

Luo Z, Viscarra-Rossel RA, Qian T (2021) Similar importance of edaphic and climatic factors for controlling soil organic carbon stocks of the world. Biogeosciences 18:2063–2073. https://doi.org/10.5194/bg-18-2063-2021

Magerl A, Le Noë J, Erb KH, Bhan M, Gingrich S (2019) A comprehensive data-based assessment of forest ecosystem carbon stocks in the US 1907–2012. Environ Res Lett 14:125015. https://doi.org/10.1088/1748-9326/ab5cb6

Mikhailova EA, Groshans GR, Post CJ, Schlautman MA, Post GC (2019) Valuation of soil organic carbon stocks in the contiguous United States based on the avoided social cost of carbon emissions. Resources 8:153. https://doi.org/10.3390/resources8030153

Minasny B, Arrouays D, McBratney AB et al (2018) Rejoinder to comments on Minasny et al., 2017 Soil carbon 4 per mille Geoderma 292, 59–86. Geoderma 309:124–129. https://doi.org/10.1016/j.geoderma.2017.05.026

Nahlik AM, Fennessy MS (2016) Carbon storage in US wetlands. Nat Commun 7:13835. https://doi.org/10.1038/ncomms13835

National Academies of Sciences, Engineering, and Medicine (2019) Negative emissions technologies and reliable sequestration: a research agenda. The National Academies Press, Washington, DC. https://doi.org/10.17226/25259NAS2019

Nave L, Marín-Spiotta E, Ontl T, Peters M, Swanston C (2019) Soil carbon management. In: Busse M, Giardina CP, Morris DM, Page-Dumroese DS (eds) Global change and forest soils,

Developments in soil science 36. Elsevier, Amsterdam, pp 215–257. https://doi.org/10.1016/B978-0-444-63998-1.00011-2

O'Neill KP, Amacher MC, Perry CH (2005) Soils as an indicator of forest health: a guide to the collection, analysis, and interpretation of soil indicator data in the Forest Inventory and Analysis program. General Technical Report NC-258. US Department of Agriculture, Forest Service, North Central Research Station, St. Paul, Minnesota, USA

Olson DM, Dinerstein E, Wikramanayake ED et al (2001) Terrestrial ecoregions of the world: a new map of life on Earth. Bioscience 51:933–938

Oswalt SN, Smith WB, Miles PD et al (2019) Forest resources of the United States, 2017: a technical document supporting the Forest Service 2020 RPA Assessment. Gen. Tech. Rep. WO-97. U.S. Department of Agriculture, Forest Service, Washington Office, Washington, DC, 223 p. https://doi.org/10.2737/WO-GTR-97

Pattison R, Andersen HE, Gray A et al (eds) (2018) Forests of the Tanana Valley State Forest and Tetlin National Wildlife Refuge Alaska: results of the 2014 pilot inventory. Gen Tech Rep PNW-GTR 967. U.S. Department of Agriculture Forest Service, Pacific Northwest Research Station, Portland, 80 p

Penuelas J, Fernández-Martínez M, Vallicrosa H et al (2020) Increasing atmospheric CO_2 concentrations correlate with declining nutritional status of European forests. Commun Biol 3:125. https://doi.org/10.1038/s42003-020-0839-y

Ping CL, Michaelson GJ, Jorgenson MT et al (2008) High stocks of soil organic carbon in the North American Arctic region. Nat Geosci 1:615–619. https://doi.org/10.1038/ngeo284

Pouyat RV, Yesilonis ID, Nowak DJ (2006) Carbon storage by urban soils in the United States. J Environ Qual 35:1566–1575. https://doi.org/10.2134/jeq2005.0215

Prăvălie R (2021) Exploring the multiple land degradation pathways across the planet. Earth Sci Rev 220:103689. https://doi.org/10.1016/j.earscirev.2021.103689

Prietzel J, Wiesmeier M (2019) A concept to optimize the accuracy of soil surface area and SOC stock quantification in mountainous landscapes. Geoderma 356:113922. https://doi.org/10.1016/j.geoderma.2019.113922

Ramcharan A, Hengl T, Nauman T et al (2018) Soil property and class maps of the conterminous United States at 100-meter spatial resolution. Soil Sci Soc Am J 82:186–201. https://doi.org/10.2136/sssaj2017.04.0122

Sanderman J, Hengl T, Fiske GJ (2017) Soil carbon debt of 12,000 years of human land use. Proc Natl Acad Sci U S A 114:9575–9580. https://doi.org/10.1073/pnas.1706103114

Saugier B, Roy J, Mooney HA (2001) Estimations of global terrestrial productivity: converging toward a single number? In: Saugier B, Roy J, Mooney HA (eds) Terrestrial global productivity. Academic, Cambridge, MA, pp 543–557

Scharenbroch B, Day D, Trammell T, Pouyat R (2018) Urban soil carbon storage. In: Lal R, Stewart BA (eds) Urban soils. Taylor & Francis, CRC Press, Boca Raton, pp 137–154

Schuur EAG, McGuire AD, Romanovsky V, Schädel C, Mack M (2018) Chapter 11: arctic and boreal carbon. In: Cavallaro N, Shrestha G, Birdsey R et al (eds) Second state of the carbon cycle report (SOCCR2): a sustained assessment report. U.S. Global Change Research Program, Washington, DC, pp 428–468. https://doi.org/10.7930/SOCCR2.2018.Ch11

Selmants PC, Giardina CP, Jacobi JD, Zhu Z (eds) (2017) Baseline and projected future carbon storage and carbon fluxes in ecosystems of Hawai'i. U.S. Geological Survey Professional Paper 1834, 134 p. https://doi.org/10.3133/pp1834

Shi Z, Allison SD, He Y et al (2020) The age distribution of global soil carbon inferred from radiocarbon measurements. Nat Geosci 13:555–559. https://doi.org/10.1038/s41561-020-0596-z

Shu S, Jain AK, Kheshgi HS (2020) Investigating wetland and nonwetland soil methane emissions and sinks across the contiguous United States using a land surface model. Glob Biogeochem Cycles 34:e2019GB006251. https://doi.org/10.1029/2019GB006251

Sleeter BM, Liu J, Daniel C et al (2018) Effects of contemporary land-use and land-cover change on the carbon balance of terrestrial ecosystems in the United States. Environ Res Lett 13:045006. https://doi.org/10.1088/1748-9326/aab540

Soudzilovskaia NA, van Bodegom PM, Terrer C et al (2019) Global mycorrhizal plant distribution linked to terrestrial carbon stocks. Nat Commun 10:5077. https://doi.org/10.1038/s41467-019-13019-2

Spencer S, Ogle AM, Breidt FJ, Goebel JJ, Paustian K (2011) Designing a national soil carbon monitoring network to support climate change policy: a case example for US agricultural lands. Greenhouse Gas Measure Manage 1(3–4):167–178. https://doi.org/10.1080/20430779.2011.637696

Sperow M (2016) Estimating carbon sequestration potential on US agricultural topsoils. Soil Tillage Res 155:390–400. https://doi.org/10.1016/j.still.2015.09.006

Sperow M (2020) Updated potential soil carbon sequestration rates on U.S. agricultural land based on the 2019 IPCC guidelines. Soil Tillage Res 204:104719. https://doi.org/10.1016/j.still.2020.104719

Sperow M, Eve M, Paustian K (2003) Potential soil C sequestration on US agricultural soils. Clim Chang 57:319–339

Sperry JS, Venturas MD, Todd HN et al (2019) The impact of rising CO_2 and acclimation on the response of US forests to global warming. Proc Natl Acad Sci U S A 116:25734–25744. https://doi.org/10.1073/pnas.1913072116

Spohn M (2020) Increasing the organic carbon stocks in mineral soils sequesters large amounts of phosphorus. Glob Chang Biol 26:4169–4177. https://doi.org/10.1111/gcb.15154

Staub B, Rosenzweig C (1992) Global Zobler soil type, soil texture, surface slope, and other properties: digital raster data on a 1-degree geographic (Lat/Long) 180×360 grid. Global Ecosystems Database Version 2.0. http://dss.ucar.edu/datasets/ds770.0/

Stott DE (2019) Recommended soil health indicators and associated laboratory procedures. Soil Health Technical Note No. 430-03. U.S. Department of Agriculture, Natural Resources Conservation Service

Sulman BN, Harden J, He Y et al (2020) Land use and land cover affect the depth distribution of soil carbon: insights from a large database of soil profiles. Front Environ Sci 8:146. https://doi.org/10.3389/fenvs.2020.00146

Sundquist ET, Ackerman KV, Bliss NB et al (2009) Rapid assessment of U.S. forest and soil organic carbon storage and forest biomass carbon sequestration capacity. U.S. Geological Survey Open-File Report 2009–1283, 15 p. http://pubs.usgs.gov/of/2009/1283/

Tifafi M, Guenet B, Hatté C (2018) Large differences in global and regional total soil carbon stock estimates based on SoilGrids, HWSD, and NCSCD: Intercomparison and evaluation based on field data from USA, England, Wales, and France. Glob Biogeochem Cycles 32:42–56. https://doi.org/10.1002/2017GB005678

Toor GS, Yang YY, Das S, Dorsey S, Felton G (2021) Soil health in agricultural ecosystems: current status and future perspectives. Adv Agron 168:157–201. https://doi.org/10.1016/bs.agron.2021.02.004

United States Environmental Protection Agency (2021) Inventory of U.S. greenhouse gas emissions and sinks: 1990–2019. EPA 430-R-21-005. U.S. Environmental Protection Agency, Washington, DC. https://www.epa.gov/ghgemissions/inventory-us-greenhouse-gas-emissions-and-sinks

Wagner G (2021) Recalculate the social cost of carbon. Nat Clim Chang 11:293–294

Waring BG, Sulman BN, Reed S et al (2020) From pools to flow: the PROMISE framework for new insights on soil carbon cycling in a changing world. Glob Chang Biol 26:6631–6643. https://doi.org/10.1111/gcb.15365hea

Wills S, Loecke T, Sequeira C et al (2014) Overview of the U.S. Rapid Carbon Assessment Project: sampling design, initial summary and uncertainty estimates. In: Hartemink A, McSweeney K (eds) Soil carbon. Progress in soil science. Springer, Cham, pp 95–104. https://doi.org/10.1007/978-3-319-04084-4_10

Woodall CW, Coulston JW, Domke GM et al (2015) The US forest carbon accounting framework: stocks and stock change, 1990–2016. US Department of Agriculture, Forest Service, Northern Research Station

Yu W, Weintraub SR, Hall SJ (2021) Climatic and geochemical controls on soil carbon at the continental scale: interactions and thresholds. Glob Biogeochem Cycles 35:e2020GB006781. https://doi.org/10.1029/2020GB006781

Zamanian K, Zhou J, Kuzyakov Y (2021) Soil carbonates: the unaccounted, irrecoverable carbon source. Geoderma 384:114817. https://doi.org/10.1016/j.geoderma.2020.114817

Zhang L, Xiao J, Zheng Y, Li S, Zhou Y (2020) Increased carbon uptake and water use efficiency in global semi-arid ecosystems. Environ Res Lett 15:034022. https://doi.org/10.1088/1748-9326/ab68ec

Zhu Z, McGuire AD (eds) (2016) Baseline and projected future carbon storage and green-house-gas fluxes in ecosystems of Alaska. U.S. Geological Survey Professional Paper 1826, 196 p. https://doi.org/10.3133/pp1826

Zomer RJ, Bossio DA, Sommer R, Verchot LV (2017) Global sequestration potential of increased organic carbon in cropland soils. Sci Rep 7:15554. https://doi.org/10.1038/s41598-017-15794-8

Chapter 3
Soil Organic Carbon Sequestration

Abstract Terrestrial biomes in the U.S. can be managed for SOC sequestration. Sequestration for climate change adaptation and mitigation occurs when the soil C inputs are derived from atmospheric carbon dioxide (CO_2) fixed by photosynthesis within a biome, and the synthesized SOC is protected and stabilized for long periods of time. Aside soil and land-use management practices, elevated CO_2, nitrogen (N) additions, warming, irrigation and increases in biomass but also natural disturbances affect SOC stocks. The SOC sequestration in managed land of forest biomes in the U.S. can be managed by practices including: (i) harvesting, (ii) thinning, (iii) fertilization, (iv) liming, (v) drainage, (vi) irrigation, (vii) tree species selection and (viii) control of understory vegetation, and by managing natural disturbances. Management of stand-replacing disturbances (i.e., fire, insect outbreaks) is particularly promising to enhance SOC sequestration. However, forest management is focused on producing timber by silviculture and, until recently, not on soil management including SOC stocks resulting in limited understanding on how to enhance forest SOC sequestration. Fire strongly affects SOC sequestration in the boreal forest/taiga biome, but it is unclear how recent changes in fire size, severity and intensity together with changes in insect and pathogen outbreaks alter SOC stocks. Importantly, it is not possible to fully control and manage SOC sequestration in boreal U.S. forests because of its scale and remoteness. In contrast, forest management interventions in the temperate coniferous forest biome during harvesting, thinning, reforestation and prescribed burning can potentially enhance SOC sequestration. Reducing the extent of harvested area on a landscape level, N-fertilization, and introduction/favoring faster-growing trees species and those more tolerant of heat or drought are among the management options. The SOC sequestration in both temperate coniferous, and broadleaf and mixed U.S. forest biomes share the same key SOC vulnerabilities associated with harvest and fire. Specifically, recent changes in fire regimes in western U.S. forests are a major concern for the fate of SOC. In the tropical forest biome, hurricanes, typhoons and cyclones may increasingly affect SOC sequestration. Otherwise, SOC sequestration in tropical forests may be enhanced by: (i) fire management, (ii) prevention of grass invasions, (iii) selection of high-SOC species for plantations, (iv) mixed-species plantations, (v) reforestation of burned areas, (vi) grazer density control, (vii) refor-

K. Lorenz, R. Lal, *Soil Organic Carbon Sequestration in Terrestrial Biomes of the United States*, https://doi.org/10.1007/978-3-030-95193-1_3

estation, (viii) facilitation of N-fixer establishment, (ix) control of soil erosion, (x) selection of high-SOC species or genetic families on degraded soils and for plantations, and (xi) retaining logging residues. The SOC sequestration in the temperate grassland, savanna, and shrubland biome in the U.S. may be enhanced by: (i) improved grazing management, (ii) fertilization, (iii) irrigation, (iv) increasing species diversity, and (v) sowing legumes and improved grass species. In contrast, management activities to increase SOC sequestration in the tundra biome are limited. Terrestrial wetlands in the U.S. are not managed for SOC sequestration. However, restoration of drained peatlands to wetlands, wetland agriculture ('paludiculture') and reduction in peat mining may contribute to SOC sequestration. The potential for management of SOC sequestration in deserts and xeric shrublands is limited as plants are often near their physiological limits for temperature and water stress. Among the opportunities to enhance SOC sequestration are restoration of degraded lands and improved grazing management. Management practices to enhance cropland SOC sequestration in the U.S. include: (i) maintaining permanent cropland cover with vegetation (i.e., elimination of summer fallow, use of perennials and cover crops), (ii) protecting the soil from erosion (i.e., reduced tillage or no-till (NT), maintaining residue cover), and (iii) improved nutrient and water management. Irrigation, and applying organic fertilizers and biochar can also contribute to SOC sequestration in U.S. croplands. Human activities, i.e., land clearing, removal of vegetation, and disturbance of soils including adding impervious cover associated with construction activities affect SOC sequestration in settlements and urban areas. However, these soils are not managed for SOC sequestration, and any recommendations on SOC-enhancing soil and land use management practices are premature. This chapter will summarize potential alterations in SOC sequestration by soil and land-use management practices, and the effects of climate and global changes on sequestration processes. The chapter will also present approaches for carbon monitoring and accounting in terrestrial ecosystems in the U.S., and how SOC sequestration in terrestrial biomes is affected by natural disturbances and how sequestration can potentially be enhanced by management interventions.

Keywords Soil organic carbon stock · Soil organic carbon sequestration · Soil organic carbon saturation · Land-use changes · Soil and land-use management · Climate change · Global change · Soil organic carbon dynamics · Carbon monitoring · Carbon accounting · Natural disturbances · Management practices

3.1 Introduction

The soil organic carbon (SOC) stocks of terrestrial biomes in the U.S. vary over time by natural and human-induced changes. In 2019, mineral soils of forests, and mineral and organic soils of croplands sequestered SOC (Table 3.1). Biome SOC stocks may differ as result of changes in the area of a terrestrial biome and/or changes in SOC stocks without changes in biome area. Primary drivers of SOC

Table 3.1 Net sequestration of organic carbon in soils by land uses in the United States in the year 2019 (negative values indicate losses; modified from U.S. EPA 2021)

Land use		Tg C	Million tn C
Forest land remaining forest land	Mineral soil	0.7	0.8
	Organic soil	−0.3	−0.3
Cropland remaining cropland	Mineral and organic soil	4.0	4.4
Grassland remaining grassland	Mineral and organic soil	−4.0	−4.4
Peatlands remaining peatlands		−0.2	−0.2
Settlements remaining settlements		−4.3	−4.7

dynamics are land use changes (Prăvălie 2021). Increases in the SOC stock of a specified biome area can contribute to SOC sequestration for climate change adaptation and mitigation when: (i) carbon dioxide (CO_2) is fixed by photosynthesis within the biome, (ii) the biomass-derived organic carbon (OC) enters the soil where it is protected and stabilized against loss for long periods of time, and (iii) no organic soil amendments are imported from outside the biome. For the climate benefit of SOC sequestration, a formal definition of SOC sequestration is the integral of an amount of carbon (C) removed from the atmosphere stored over the time horizon it remains within a soil (Sierra et al. 2021). There may be a broad U.S. public support for soil C storage as a climate change mitigation strategy (Sweet et al. 2021).

Inputs of biomass C derived from plant photosynthesis will increase SOC sequestration as long as soils are in disequilibrium – until a new balance between carbon inputs and outputs is reached. Soils with lower initial SOC stocks, and with lower initial C stocks in particulate and mineral-associated organic matter (POM and MAOM, respectively), and in pyrogenic OM may have a larger potential for sequestering C following changes in soil and land-use management, particularly in agroecosystems, than those with higher total SOC and fraction SOC stocks (Luo et al. 2020). Globally, a major potential for C sequestration is in cropland soils, especially those with large yield gaps and/or large historic SOC losses (Amelung et al. 2020). The most effective way to accumulate SOC is to increase organic C inputs as SOC is generally driven by vegetation productivity (Todd-Brown et al. 2013). However, meaningful increases in C sequestration at one farm or region must occur without simultaneous reductions in SOC at another location from where the organic material is transported from. Thus, organic C inputs into soil must be produced on-site, e.g., by enhancing crop production and green manure (Amelung et al. 2020).

Efforts to manage soils for SOC sequestration depend also on apparent SOC saturation (Craig et al. 2021). Soil minerals, in particular, can protect OM from degradation (Kleber et al. 2021). If mineral saturation is the primary cause of SOC saturation in a soil, then efforts to increase SOC may focus on: (i) augmenting particulate SOC (e.g., through recalcitrant inputs), (ii) promoting SOC translocation to deeper subsurface horizons where SOC concentrations are farther from saturation (e.g., by planting deeply rooted species), or (iii) amending soils with reactive minerals (Craig et al. 2021). The reactive mineral/metal-associated C pools from many forest and grassland ecosystems in the U.S. are possibly not yet saturated (Yu et al.

2021c). These soils may have additional capacity for protection of SOC by minerals or metals (Yu et al. 2021c). If microbial constraints are limiting mineral-associated SOC buildup, then management efforts could focus on investigating and alleviating constraints on microbial biomass (e.g., reducing soil compaction or acidification). If SOC stocks are primarily regulated by microbial traits, the characterization and management of the microbial community may promote SOC storage (Craig et al. 2021). Overall, there is indication that some soils in the U.S. are below the apparent saturation point of mineral-associated OM (Yu et al. 2021c).

3.2 Temporal Changes in Land-Use

In the long-term, the total cropland area in the conterminous U.S. (CONUS, i.e., the 48 adjoining U.S. states on the continent of North America), Alaska and Hawaii decreased by about 18% between 1949 and 2012 (Bigelow and Borchers 2017). Between 2008 and 2012, however, nearly 30 thousand km^2 (12 thousand mi^2) of U.S. cropland expansion occurred where new croplands primarily replaced grasslands (Spawn et al. 2018). This conversion reduced SOC stocks to 1-m (39-in) depth by 30.4%. On average, 55 Mg C ha^{-1} (25 tn C ac^{-1}) were released form above and belowground biomass C and SOC stocks in 2008–2012 (Spawn et al. 2018). From 2012 to 2015, cropland acreage in CONUS, Hawaii, Puerto Rico, and the U.S. Virgin Islands further increased by about 0.020 million km^2 (0.007 million mi^2), with most of the gain from land converted from Conservation Reserve Program (CRP) land (USDA 2018).

Temperate grassland, pasture and range land use in CONUS, Alaska and Hawaii increased through the early 1960s but then entered a long-term decline (Bigelow and Borchers 2017). From 2007 to 2012, the temperate grassland, pasture and range area increased again. Forest-use area declined from 1949 to 1997, increased from 1997 to 2007, and declined again over 2007–2012. Data collected 2000–2017 indicated that change in land use from temperate forests is more likely with increasing human population and housing growth rates (Fitts et al. 2021). Further, areas closer to cities and coastal areas showed a higher risk of transition to non-forests. During 1945–2012, both special-use and urban areas expanded in CONUS, Alaska and Hawaii (Bigelow and Borchers 2017). For example, about 0.170 million km^2 (0.067 million mi^2) of land was newly developed in CONUS, Hawaii, Puerto Rico, and the U.S. Virgin Islands between 1982 and 2015, representing a 60% increase in the total developed land area (USDA 2018). Further, land under miscellaneous other uses in CONUS, Alaska and Hawaii declined over 1974–2012 (Bigelow and Borchers 2017).

Detailed land use (National Land Use Database, NLUD) and land cover (National Land Cover Database, NLCD) datasets provide insights into how land cover in the U.S. is changing over time (Wentland et al. 2020). Homer et al. (2020) assessed land change patterns across CONUS from 2001 to 2016 based on the 2016 NLCD product suite. Deciduous and Conifer Forest NLCD Class area decreased from 780,529 and 963,379 km^2 (301,364 and 371,963 mi^2) in 2001 to 756,813 and 923,780 km^2 (292,207 and 356,673 mi^2) in 2016, respectively. Decreases in extent during this period were also reported for Pasture/Hay which decreased from 551,345 km^2 (212,875 mi^2) to 507,568 km^2 (195,973 mi^2). In contrast, increases in land cover extent occurred for Cultivated Crops and Grassland Herbaceous covering 1,264,559 and 1,092,991 km^2 (488,249 and 422,006 mi^2) in 2001, and 1,313,114 and 1,118,412 km^2 (506,996 and 431,821 mi^2) in 2016, respectively. Further, extent of all developed NLCD Classes increased over 15 years showing most persistent and permanent land change increases. For example, strong net increases were reported for Developed – Medium Density covering 45,991 km^2 (17,757 mi^2) in 2001 and 56,283 km^2 (21,731 mi^2) in 2016 (Homer et al. 2020).

Significant land-use changes in the U.S. may continue to occur in the future. For example, Lawler et al. (2014) projected a large increase in cropland area (282,000 km^2 [88,000 mi^2]) under a scenario with high crop demand, compared with another scenario with a loss of cropland (112,000 km^2 [43,000 mi^2]) for CONUS until the year 2051. Major land-use changes including land abandonment and land-use expansion were widespread in five spatially explicit scenarios of future land-use changes in 2001–2051 (730,000–950,000 km^2 [282,000–367,000 mi^2]), especially in the eastern U.S. and along the West Coast (Martinuzzi et al. 2015). The projected land-use changes will affect SOC sequestration and stocks in the U.S. (Lawler et al. 2014).

This chapter summarizes potential alterations in SOC sequestration by soil and land-use management practices, and the effects of climate and global changes on sequestration processes. The chapter also presents approaches for carbon monitoring and accounting in terrestrial ecosystems in the U.S., and how SOC se-questration in terrestrial biomes is affected by natural disturbances and how se-questration can potentially be enhanced by management interventions.

3.3 Alterations in Soil Organic Carbon Sequestration

3.3.1 Soil and Land-Use Management

The SOC sequestration in terrestrial biomes of the U.S. can result from management-induced changes in species composition and abundance, primary productivity, land-surface hydrology, and the biogeochemical cycles of C, nitrogen (N), and phosphorus (P; Ellis and Ramankutty 2008). The changes in SOC stocks after land-use change (LUC) in terrestrial biomes depend also on socioeconomic in addition to biophysical factors (Duarte-Guardia et al. 2020). Globally, socioeconomic variables such as indices of poverty, population growth, and levels of corruption are important for explaining SOC stock changes after LUC. However, gaps in the global soil profile dataset still exist for western North America, and socioeconomic variables are generally characterized by coarse spatial resolution (Duarte-Guardia et al. 2020). Nevertheless, the social context needs also to be actively considered for managing SOC stocks in terrestrial biomes of the U.S.

Nitrogen addition resulted in increases in SOC stocks globally by 11.0% estimated based on a data set spanning 60 years across 369 sites, and the SOC response to N addition increased with N addition duration (Xu et al. 2021b). The positive effects of N addition on SOC were independent of ecosystem types, mean annual temperature and precipitation. The SOC increases largely resulted from the enhanced plant C input to soils coupled with reduced C loss from decomposition, and amplification was associated with reduced microbial biomass and respiration under long-term N addition. Thus, N addition will enhance SOC sequestration over time and may contribute to future climate change mitigation (Xu et al. 2021b). In addition to N, SOC sequestration in croplands may be promoted by addressing nutrient mining by modern crop cultivars (Chaplot 2021). Decades of agriculture intensification have exported large amounts of nutrients (N, P, potassium [K] but also sulfur [S], calcium [Ca], magnesium [Mg], iron [Fe], chlorine [Cl], manganese [Mn], zinc [Zn], copper [Cu], boron [B], molybdenum [Mo]) from soil. Any nutrient deficiency in the soil is likely to induce mining of soil organic matter (SOM) by plants leading to OM decomposition and SOC loss (Chaplot 2021).

Agroecosystems

Improvements in land management for crops and grazing can be used specifically to limit warming by reducing GHG emissions as well as by sequestering C (Girardin et al. 2021). Agroecosystems can be particularly managed for SOC sequestration through increases in SOC stocks, e.g., by improving cropland and grazing land management, and restoring degraded lands and cultivated organic soils (Smith et al. 2008). By both increasing C inputs and reducing SOC losses, SOC sequestration can be improved (Tiefenbacher et al. 2021). Practices include for example: (i) management of water including irrigation in arid and semi-arid regions; (ii) land use change to an agroecosystem with a higher equilibrium SOC stock; (iii) set-aside; (iv) agroforestry; (v) livestock management; and (vi) manure management. In 2015, 257 thousand km^2 (99 thousand mi^2) of agricultural land were irrigated, consuming 277 billion L (73 billion gal) of water per day (Dieter et al. 2018). Particularly in surface soils of arid and semi-arid regions, effect of irrigation on plant growth outweighs the effects on SOC decay by microbes (Emde et al. 2021). Globally, irrigated agriculture tends to increase SOC stocks by 5.9% based on a meta-analysis of studies that compared SOC in the same irrigated plots or fields at the beginning and end of an experiment. However, irrigation-induced changes in SOC stocks varied by climate and soil depth, with the greatest increase in SOC observed at irrigated semi-arid sites at the 0–10 cm (0–3.9 in) depth (14.8%). Overall, irrigated agriculture tends to increase SOC stocks, particularly in surface soils, in fine-to medium-textured soils, in arid to semi-arid climates and under sprinkler irrigation (Emde et al. 2021).

Practices that may increase SOC stocks in croplands include: (i) management of vegetation, including high-input C practices, e.g., improved rotations, cover crops, and perennial cropping systems; (ii) nutrient management to increase plant C returns to the soil, e.g., through optimized fertilizer application rate, type, timing, and precision application; (iii) reduced tillage intensity and residue retention; and (iv) improved water management (Smith 2016). Crop management measures (rotation, cover cropping, residue retention), soil and water measures (irrigation, tillage), soil amendments (enhanced efficiency, biochar), fertilizer use (organic, mineral, combined organic-mineral) and "4R'" fertilizer strategies (right source, rate, timing, placement) all positively impact SOC (Young et al. 2021). The largest effect on cropland SOC among these measures was due to biochar, followed by organic fertilizer input.

No-till (NT) was not originally designed as a strategy to sequester SOC, but to reduce soil erosion and degradation (Lal et al. 2007). Severe water erosion affects ~760,000 km^2 (293,000 mi^2) in the U.S. (Prăvălie 2021). Thus, enhancing both cover crops and crop rotation complexity may be needed to boost SOC sequestration performance of NT (Blanco-Canqui 2021). For example, compared to tilled soils, the higher SOC stock of NT of +4.7 Mg C ha^{-1} (2.1 tn C ac^{-1}) in the 0–60 cm (0–24 in) depth with an average of 11 years of NT depended on the association of NT with increased crop frequency and the inclusion of legume cover crops (Nicoloso and Rice 2021).

The SOC sequestration potential of relatively new management practices, such as the application of inorganic C or biochar and agroforestry, is also uncertain because published data from long-term field experiments are lacking (Tiefenbacher et al. 2021).

Agricultural practices that have been demonstrated to reduce greenhouse gas (GHG) emissions and increase SOC sequestration in the U.S. include cover cropping, more varied crop rotations, agroforestry and silvopasture, perennial crops, rotational grazing, and dry manure management (Lehner and Rosenberg 2019). Because C and N cycles are tightly coupled in soils, the build-up of SOC in agroecosystems may enhance primary production and enhance further SOC storage, but also increase the risk of nitrous oxide (N_2O) emissions because of the increase in N sources and the shift to soil environmental conditions more favorable to N_2O emissions (Guenet et al. 2021). In the long term, increased N_2O emissions by SOC sequestering practices may offset some of the climate benefits of increased SOC stocks (Lugato et al. 2018).

Conservation agriculture (CA) practices such as: (i) soil mulch cover on the soil surface, (ii) continuous minimum soil disturbance, and (iii) diversification of crop species can lead to increases in SOC stocks (Kassam et al. 2019). Only the concomitant implementation of all three practices constitutes a CA system. In general, CA practices extend soil lifespans and may promote soil thickening, increasing the potential for C storage (Evans et al. 2020). About 432,000 km^2 (167,000 mi^2) of 'arable' cropland in the U.S. were under CA practices in 2015/2016, increasing from 265,000 km^2 (102,000 mi^2) in 2008/2009 (Kassam et al. 2019). Increases in SOC stocks at the soil surface are observed particularly in regions where soil and climatic conditions are favorable for biomass production, and where the CA system does not negatively impact yield (Page et al. 2020). For example, increased SOC stocks and increases yields under CA are most likely in warm and arid regions (Sun et al. 2020). CA is likely to increase SOC stocks in mid and eastern parts of the U.S., and in California. Otherwise, in semiarid to humid regions, CA may increase SOC stocks with no effect on crop yield. In cold humid and tropical humid climates, however, CA may likely result in both reduced SOC stocks and crop yields (Sun et al. 2020).

In addition to CA, regenerative agricultural practices may also have the potential to increases SOC stocks and persistence. For example, Lehmann et al. (2020) proposed regenerative SOC practices consistent with the promotion of functional diversity to increase SOC persistence. Practices include mixtures of inputs, and a diversity of plant species to stimulate a diverse microbial community and rhizodeposits. Increasing crop plant diversity from monocultures over two- to four-species mixtures may also increase annual primary productivity, resulting in overall higher plant biomass (Chen et al. 2021). Further, reduced soil mixing (by tillage), should be explored to increase SOC persistence and sequestration. It is important to better understand how to sequester SOC by increasing persistence based on functional complexity in comparison to merely increasing organic C inputs (Lehmann et al. 2020). For example, based on a second-order meta-analysis, Tamburini et al. (2020) reported that agricultural diversification marginally enhanced C sequestration (storage), in particular, by the diversification practices organic amendment, reduced tillage, and inoculation (e.g., arbuscular mycorrhizal (AM) fungi, N_2-fixing bacteria, and growth-promoting bacteria).

Maintaining land cover with vegetation (especially with deep-rooted perennials and cover crops), and protecting the soil from erosion can increase SOC stocks (Hristov et al. 2018). Deep-rooted perennial crops are important for stabilization and storage of SOC over long time scales in deep soil (Peixoto et al. 2020). Recommended practices in grazing land management/pasture improvement include nutrient (i.e., P) management, increased productivity (e.g., fertilization), fire management and species introduction (including legumes; Smith et al. 2008). The management of organic soils (e.g., avoided drainage of wetlands) positively affects SOC stocks. Increases in SOC stocks may also occur by the restoration of degraded lands (e.g., erosion control, organic and nutrient amendments). Further, more efficient use of manure and biosolids as nutrient sources can have positive effects on SOC stocks. Bioenergy (e.g., energy crops, solid, liquid, biogas, residues) can also contribute to an increase in SOC stocks (Smith et al. 2008).

Forest Ecosystems

The SOC sequestered, stored and cycled under forests is poorly understood due to its complexity in mechanisms of storage and inaccessibility at depth (Price et al. 2012). Fluxes of SOC in forests are often ignored regarding their mitigation potential although the contribution from forest pathways to SOC sequestration may be substantial (Bossio et al. 2020). Management of forests can be used specifically to limit warming (Girardin et al. 2021). Natural climate solutions, subset of nature-based solutions, aim to reduce atmospheric GHG concentrations. This includes avoiding emissions by protecting forest ecosystems, i.e., limiting deforestation, and, thus, reducing C release. Another way is to improve land management for timber to reduce GHG emissions as well as to sequester C (Girardin et al. 2021). Globally, conservation and restoration of tropical forests is a priority for climate change mitigation as these forests tend to have both the highest rates of C sequestration in young stands, fueled by their generally high C flux rates, and the highest mean biomass (Anderson-Teixeira et al. 2021). Forest conservation and restoration efforts should also include high-biomass old-growth temperate forests. Overall, the conservation of mature forests will yield greater climate benefits compared to young forests as their high C stocks if lost through disturbance may not be recovered on time scales most relevant to avoiding dangerous climate change (Anderson-Teixeira et al. 2021).

Forest management can change SOC dynamics by altering: (i) infiltration of dissolved organic carbon (DOC) from litter into the mineral soil, (ii) fine root turnover and/or exudation, (iii) bacterial and fungal community composition, (iv) particulate decomposition from forest fires, and (v) stabilization by biochemical resistance, physical protection, poor drainage, fire and deep charcoal burial (Price et al. 2012). Large knowledge gaps exist for fine root turnover, specifically, for tropical forests as aboveground biomass is predominantly studied (Iversen et al. 2017). Overall, the potential for additional SOC storage from improved management practices in natural and plantation forests is uncertain (Bossio et al. 2020).

Losses from SOC stocks of U.S. forests can potentially be minimized, and SOC stocks increased by forest management including increased reforestation, improved forest management and avoided deforestation. In general, SOC stocks at or near the soil surface are vulnerable to disturbances that alter the processes responsible for maintenance of those stocks, and management that restores or maintains those processes can often reverse SOC losses over decadal timescales (Nave et al. 2019a). The subsoil SOC stock is typically less vulnerable to short-term losses, but the factors that promote this stability are difficult to manipulate in a positive direction through management. The SOC is often not managed directly but through its interactions with other components of the forest ecosystem, such as the vegetation, according to constraints imparted by fundamental bottom-up factors such as parent material and soil characteristics (Nave et al. 2019a).

Sequestration of SOC in managed forests in the U.S. includes practices such as thinning, drainage, elongation of the rotation period, fertilization, liming, site preparation, management of fire, windthrow and insects, afforestation and reforestation, harvest management and input of harvest residues (Lorenz and Lal 2010). A change in tree species composition, in particular, is a promising approach to enhance SOC sequestration. Well-replicated common garden experiments allow the effects of tree species to be separated from other confounding factors. These experiments indicate that tree species differ in aboveground C sequestration rates, the size of the SOC stock and its stabilization (Angst et al. 2019; Binkley and Fisher 2020; Verheyen et al. 2016). For example, softwood forest types in CONUS currently exhibit a higher rate of increase in the amount of C in aboveground live tree biomass (Hoover and Smith 2021). Tree-mediated shifts to soil properties are linked to mycorrhizal association type which has implications for predicting changes in biogeochemical processes following shifts in tree species composition (Yates et al. 2021). However, observations at non-experimental forest sites are less clear (Vesterdal et al. 2013). Based on a global meta-analysis, Boča et al. (2014) reported that forest floor C stocks were 38% higher under conifers, mineral SOC stocks were similar, and whole soil SOC stocks were 14% higher under conifers. However, the quantity of available data was not large, and the methodologies used were diverse. Thus, any conclusions drawn must be regarded as working hypotheses from which to design future targeted investigations that expand the database (Boča et al. 2014). Overall, productive forests with a high rate of aboveground and belowground litterfall circulate a large amount of C, and are a precondition for efficient SOC sequestration (Jandl et al. 2007). Thus, optimized forest management with regard to SOC sequestration should aim to secure a high productivity of the forest on the C input side, and avoid soil disturbances as much as possible on the C output side (Jandl et al. 2007).

Wetlands and Peatlands

Wetlands including peatlands in the U.S. are generally C sinks, and the management objective for C sequestration in organic soils is to reduce SOC losses (Amelung et al. 2020). The SOC stocks of organic soils can also be increased by implementing targeted management practices (Kolka et al. 2018). For example, omitting seasonal drainage in moist soil regimes (Drexler et al. 2013), replanting with native wetland species (Bossio et al. 2020), and deeply flooding are among the options to increase wetlands SOC stocks. Restoration of wetlands, so that they sequester C, is a natural climate solution (Girardin et al. 2021). Terminating peat mining for fuel or horticultural purposes contributes to higher peatland equilibrium SOC stocks. Water levels in production agriculture wetlands can be adjusted, and managed to increase SOC stocks (Kolka et al. 2018). Further, temporarily ponded wetlands that dry down during the growing season should not be tilled and farmed carefully to not increase SOC decomposition rates. Restoration of peatlands used for agriculture to vegetated wetlands can contribute to increases in SOC stocks by: (i) decreases in CO_2 fluxes related to the oxidation of SOC while in crop production, (ii) decreases in the use of N fertilizers, (iii) decreases in lime application amendments, and (iv) increases in C sequestered in perennial vegetation (ICF International 2013). Wetland restoration including rewetting can lead to increases in SOC stocks by the opposite environmental effects of drainage (Lucchese et al. 2010).

Urban Ecosystems

A variety of practices have been proposed to increase SOC stocks of urban soils (Lorenz and Lal 2015). For examples, SOC stocks of bare urban soils can be increased by revegetation, and those of urban green space soils by fertilizing, irrigation, reduced soil disturbance and residue management (e.g., returning grass clippings). During construction activities, urban soils can potentially be improved toward SOC accumulation by adding organic amendments such as biosolids, and yard and food wastes (Brown et al. 2012). Garden management practices such as the addition of composts, mulches, mineral fertilizers and water, and cultivation of trees and shrubs, may contribute to greater SOC stocks in urban gardens (Edmondson et al. 2012). Tree planting and management of existing tree cover likely provide the greatest scope for increasing urban SOC stocks (Renforth et al. 2011). Removing soil sealing, i.e., the permanent covering of soil by completely or partly impermeable material, and promoting soil C inputs in un-sealed urban areas can also potentially result in increases in urban SOC stocks (Lorenz and Lal 2017).

3.3.2 *Climate and Global Changes*

Earth's climate is warming as the atmospheric CO_2 concentration has increased from ~277 ppm in 1750 to 412.5 ppm in 2020 (Blunden and Boyer 2020; Cui et al. 2020). In 2019, atmospheric CO_2 concentrations were higher than at any time in at least 2 million years, and concentrations of CH_4 and N_2O were higher than at any time in at least 800,000 years (IPCC 2021). Mainly due to the increased GHG concentrations, human-caused radiative forcing of 2.72 W m^{-2} in 2019 relative to 1750 has warmed the climate system.

Experimental warming increases fine-root biomass and production (Wang et al. 2021b). The C inputs from fine roots are the dominant inputs to SOC stocks compared with aboveground litter-derived C inputs. Above- and belowground litter production depends on primary production, and this is sensitive to climate variability in CONUS (Maurer et al. 2020). Annual primary production values over a 16-year period have been shown to be most sensitive to precipitation and aridity in dryland and temperate grassland ecosystems. Measurements at climate stations indicate that aridity and climatic variability is increasing in many regions in these ecosystems. Long-term drought trajectories indicate that chronic drought will increase in southern dryland areas in the U.S., and decrease in the north (Bradford et al. 2020). However, projected increases in hot-dry stress in northern drylands may also alter C cycling (Bradford et al. 2020). Dryland ecosystems in the western U.S. may be particularly vulnerable to reductions in primary production and consequent degradation of ecosystem services including C sequestration (Maurer et al. 2020). Otherwise, drought effects on wet soils in inland wetlands and peatlands may lead to an increase in oxidation of SOC (Stirling et al. 2020). Peatland and wetland soils may not easily recover between severe droughts and instead enter alternative stable states. However, understanding of the effects of drought on wet soils is incomplete (Stirling et al. 2020).

Carbon Dioxide Fertilization

The increase in atmospheric CO_2 concentration contributes to an increase in temperature but may also contribute to an increase in the land CO_2 sink. For example, from 2000 to 2014, the atmospheric CO_2 concentration increased by 29 ppm, and this may have enhanced the gross primary production (GPP) of C3 vegetation by 1.2% of global GPP or 1.8 Pg C y^{-1} (2.0 billion tn C y^{-1}; Ueyama et al. 2020). Globally, the CO_2 fertilization effect on photosynthesis has been estimated at 30% since 1900 (Haverd et al. 2020). The rising CO_2 was the largest contributor to the modelled GPP increase during 1982–2016 (Cai and Prentice 2020). In contrast, Wang et al. (2020b) reported that the global CO_2 fertilization effect has declined across most terrestrial regions of the globe from 1982 to 2015, correlating well with changing nutrient concentrations and availability of soil water (e.g., Joo et al. 2020). Further, the mean temperature of the warmest quarter has passed the thermal maximum for photosynthesis during the past decade (Duffy et al. 2021). Plant respiration rates continue to rise in contrast to sharply declining rates of photosynthesis at higher temperatures. Thus, a near halving of the land C sink strength may occur by as early as 2040 under business-as-usual emissions (Duffy et al. 2021).

Plant responses to rising atmospheric CO_2 concentrations has largely been studied by single generation exposure to elevated CO_2 but decadal multigenerational exposure studies indicate that plants may adapt to higher atmospheric CO_2 concentrations which may reduce plant biomass increases (Saban et al. 2020). Nevertheless, the single and interactive effects among global change factors on biomass allocation alters plant-and microbial-derived soil C inputs and, thus, SOC stocks. Further, ecosystems continuously adapt to interacting environmental drivers that change over time (Krause et al. 2020). Thus, the C balance of terrestrial ecosystems may currently still be affected by past anthropogenic disturbances (e.g., deforestation) and other environmental changes (e.g., climate change). These legacy effects over the twenty-first century may be much larger than the transient effects of future environmental changes. The fact that ecosystems are presently not in equilibrium with the environment needs to be considered when interpreting changes in the terrestrial C cycle (Krause et al. 2020).

For SW North America, negative GPP trends were modelled for 1982–2016, probably due to decreased precipitation or intensified dry seasons (Cai and Prentice 2020). In the future, GPP may also decrease in Eastern U.S. as in some regions significant soil moisture depletion in 0–50 cm (0–20 in) depth has already commenced or been experienced (Joo et al. 2020). Similarly, reduced soil moisture and plant available water in the late growing season in arctic and boreal ecosystems are caused by increased rate and duration of evapotranspiration due to longer growing seasons and earlier observed photosynthesis from climate warming (Madani et al. 2020). This will reduce the C uptake capacity of these biomes over the next century. Currently, high-latitude northern forests are persistent C sinks because of increases in growing season mean temperature (Liu et al. 2020b; Madani et al. 2020). In contrast, CO_2 fertilization may not be a major contributor to C exchange in boreal forests (Liu et al. 2020b). Growth enhancement in boreal forests caused by CO_2 fertilization and climate warming may be modest or non-existent (Wang et al. 2021a). However, the CO_2 fertilization effect has increased in other regions in North America (Wang et al. 2020b), such as the Great Plains in the U.S. likely driven by an intensification of management in croplands (i.e., irrigation and fertilization) or related to increasing atmospheric nutrients deposition in recent years.

Terrestrial biomes do not remain static when the climate changes rapidly (Wang et al. 2020a). Specifically, the boreal forest and tundra biomes are not resilient to changing climates as they shift to the unstable states of open woodlands. This may affect SOC sequestration and stocks. Similarly, climate-change induced increases in greening occurred in the northern grassland biome of the U.S. in the past two decades (Brookshire et al. 2020), and this may also affect SOC. The effects of elevated atmospheric CO_2 on terrestrial C cycles depend particularly on the concurrent effects of CO_2 on ecosystem function, the availability of other limiting resources, and changes in plant community diversity and composition (Fay et al. 2021). Surprisingly, SOC stocks did increase with elevated CO_2 in grasslands (+8%) but not in forests (0% change) based on a synthesis of data from 108 experiments (Terrer et al. 2021). Otherwise, plant biomass in grasslands increased less (9%) than in forests (23%). Thus, when plant biomass is strongly stimulated by elevated CO_2, SOC stocks decline. In contrast, when biomass is weakly stimulated, SOC stocks increase. This trade-off may be related to plant nutrient acquisition, in which plants increase their biomass by mining the soil for nutrients, which decreases SOC stocks (Terrer et al. 2021).

Global Changes

The combination of climate change and other changes linked to human activities is described as global change (Gauthier et al. 2015). With the global change factors elevated CO_2, N addition, warming, and irrigation, aboveground biomass increases (Zhou et al. 2020). Increases in aboveground biomass are also observed with the combinations: (i) elevated CO_2 with either N addition or warming, (ii) N addition with warming, (iii) elevated CO_2 with N addition and warming, (iv) irrigation with elevated CO_2 or N addition or warming, (v) irrigation with elevated CO_2 and warming, (vi) irrigation with N addition and warming, and (vii) irrigation with elevated CO_2 and N addition and warming. In contrast, belowground biomass is stimulated only by: (i) elevated CO_2, (ii) N addition, (iii) warming, (iv) elevated CO_2 with either N addition or warming, (v) N addition with warming, and (vi) irrigation with either elevated CO_2 or warming. Further, drought decreased both total and aboveground biomass including that of soil microbial biomass while soil microorganisms acclimate to long-term elevated precipitation (Xu et al. 2020a, b). Among the single-factor impacts on plant biomass accumulation, the effects of N addition were larger than effects of increased precipitation and elevated CO_2 (Zhou et al. 2020).

In 2007–2019, warming has had a positive impact on net CO_2 uptake during the early crop growth stage in the U.S Corn Belt, but has reduced net CO_2 uptake in both croplands and natural ecosystems during the peak growing season (Yu et al. 2021a). Future increase in summer temperature is also projected to reduce annual CO_2 sequestration in the Corn Belt by 10–20%. Thus, warming may not continue to favor CO_2 sequestration in northern mid-latitude ecosystems (Yu et al. 2021a). Worryingly, U.S. corn and soybeans are maladapted to a changing climate (Yu et al. 2021b). On the positive side, both crops have become more heat and drought tolerant over 1951–2017. However, these improvements have been achieved at the cost of reducing crop productivity under normal growing degree days and normal precipitation conditions (Yu et al. 2021b).

Beyond warming, other global change factors such as increases in precipitation, atmospheric CO_2 concentration, N deposition, and an increase in the growing season length may all contribute to an increase in SOC stocks in the U.S. (Huntzinger et al. 2017). Globally, the rising atmospheric CO_2 concentration is the dominant driver of the estimated 31% increase in terrestrial GPP since 1900 (Haverd et al. 2020). Carbon fixation by photosynthesis or GPP is the main C source for SOC formation under natural conditions. In contrast, decreases in precipitation, and emerging N and P limitations can act to constrain terrestrial C uptake and SOC formation. For example, forest regrowth following historical clearing plays a substantial role in determining the size of the forest C sink, but studies suggest sizeable contributions from growth enhancements such as CO_2 fertilization, N deposition, or climate trends supporting accelerated growth (Domke et al. 2018). Forest vegetation dynamics are already strongly influenced by global change (McDowell et al. 2020). For example, functional composition of forests across Canada shifted toward fast-growing deciduous broadleaved trees and higher drought tolerance over 65 years (Hisano et al. 2021). Further, the functional composition of colder regions shifted toward drought tolerance more rapidly with rising CO_2 than warmer regions, suggesting the greater vulnerability of boreal forests than temperate forests under ongoing global environmental changes (Hisano et al. 2021). Ultimately, the C uptake and release by forests results from the antagonistic process of rising atmospheric CO_2 and forest recovery from land-use changes, which enhance the C sink, and rising vapor pressure deficit and disturbances that reduce the C sink. Globally, forests are increasingly consisting of younger, shorter stands (McDowell et al. 2020). However, the net effect of global changes on forest SOC stocks is uncertain.

Turnover of Soil Organic Carbon
Globally, land use change and climate change are major drivers of the acceleration of turnover of SOC with a reduction in residence time by 9 years between the 1860s and the 2000s (Wu et al. 2020). Other than by management-induced changes, SOC stocks are directly and indirectly altered by climate change. Specifically, the SOC stock is globally vulnerable to perturbations by future temperature and precipitation increase (Eglinton et al. 2021). Soil warming keeps pace with air warming except where snow and ice occur (Phillips 2020). Soil temperatures at depths of 10, 20, and 50 cm (4, 8 and 20 in) in CONUS have increased on average by between 0.2 and 0.4 °C from 1948 to 2008 (Hao et al. 2014). Over the twenty-first century, rapid and deep soil warming is predicted for Mollisols under temperate grassland in the U.S. (Soong et al. 2020). Soil warming may accelerate SOC loss of whole-soil profiles primarily by stimulating the decomposer activities of microorganisms and that of SOC-degrading enzymes (Davidson and Janssens 2006; Melillo et al. 2017; Pries et al. 2017). For example, at a mixed-conifer temperate forest in the U.S., 33% of the subsoil SOC in 20–90 cm (8–35 in) depth was lost after 4.5 years of whole-soil warming (4 °C [7.2 °F]; Soong et al. 2021). This loss was primarily from unprotected POM, and warming also stimulated a sustained 30% increase in soil CO_2 efflux due to increased CO_2 production through the whole-soil profile (Soong et al. 2021).

Globally, substantial SOC losses of 232 Pg C (256 billion tn C) to 1-m (3.3-ft) depth due to 2 °C (3.6 °F) global warming have been projected even in the absence of losses of deeper permafrost C (Varney et al. 2020). García-Palacios et al. (2021) suggested that net global SOC losses with warming may occur via increases in soil microbial metabolic activity but there is no consensus on the loss magnitude. For example, SOC formation may also be enhanced at higher soil temperatures by an increase in plant productivity as the higher plant-derived soil C input under warmer conditions may offset SOC losses (Chen et al. 2020b). In fact, experimental canopy air and soil warming did not result in changes in SOC stocks globally (Liu et al. 2020c). However, knowledge of the offsets and spatiotemporal changes in variability between soil-level and free-air temperatures needs to be improved (Lembrechts et al. 2020).

The fate of the SOC stock in the U.S. under higher temperatures is also uncertain (van Gestel et al. 2018). Important advances in understanding the temperature response of the processes that control substrate availability, depolymerization, microbial efficiency, and enzyme production will be needed to predict the fate of SOC in a warmer world (Conant et al. 2011). Separating SOC into POM and MAOM aids in the understanding of its vulnerability to climate change and identification of C sequestration strategies (Lugato et al. 2021). For example, arable and coniferous forest soils (0–20 cm [0–8 in]) contain the largest and most vulnerable SOC stocks when cumulated at the European scale.

Based on data from nonagricultural ecosystems without experimental manipulation, soil respiration increased in 1987–1999 globally but became unchanged in 2000–2016, which were related to complex temporal variations of temperature anomalies and SOC stocks (Lei et al. 2021). Reductions in SOC stocks and, in particular, the chemical recalcitrant SOC with warming have been shown to be associated with increased ratios of ligninase to cellulase enzyme activities (Chen et al. 2020b). This shift in enzyme activity may also increase microbial accessibility of litter and SOC, leading to accelerated SOC loss with prolonged warming. Otherwise, less SOC loss may occur by microbially mediated dampening effects of warming on heterotrophic soil respiration (Guo et al. 2020). Thus, the potential positive feedback of soil microbial respiration in response to warming may be less than previously predicted. However, it is unlikely for heterotrophic soil respiration to fully acclimate to warming since depletion of labile C pools in soils will irreversibly change microbial community composition, shift microbial C use efficiency, and reduce microbial biomass (Lei et al. 2021).

Soil Processes

Aside increased soil temperature, soil-moisture variability may reduce the present land C sink in the U.S. (Green et al. 2019). Specifically, soil-moisture variability resulted in large modelled net biome productivity (NBP) reductions in seasonally dry climates in the western U.S. during 1971–2000. Globally, the NBP or net C accumulation by biomes is controlled by temperature, radiation and soil moisture content (Marcolla et al. 2020). In the U.S., soil moisture content shows an increasing control on the variability of NBP. In 2056–2085, negative impacts of soil moisture on mean NBP are projected to remain strong in humid climates in the south-eastern U.S. (Green et al. 2019). Thus, reductions in NBP caused by soil-moisture variability may eventually result in decreased SOC stocks in the western and south-eastern U.S.

Detrimental effects of climate change on crops, forests and livestock are increasing in the U.S. (Walthall et al. 2013). Many soil processes including erosion, compaction, acidification, salinization, toxification, and net loss of SOC are particularly sensitive to changing climate conditions. Soil erosion and land use, and SOM changes best detect agricultural responses to climate change (Hatfield et al. 2020). Regional Climate Hubs assist farmers, ranchers, forest landowners, resource managers, and rural communities in the US. to adapt to a changing climate (Steele and Hatfield 2018). Increasing daytime and/or nighttime temperatures, growing precipitation variability and intensity, shifting growing seasons, and extreme events are forcing farmers, ranchers, and foresters to consider alternative management practices. Adaptation strategies include also investing in equipment to keep confined livestock cool and hydrated, identifying better region and climate-adapted cattle (*Bos taurus* Linnaeus, 1758) breeds, crop cultivars/tree species, or adopting new management practices (Steele and Hatfield 2018).

Disturbance Regimes

Over the past four decades, burned area from wildfires has quadrupled in the U.S., increasing at a rate of 700 km^2 (270 mi^2) per year (Burke et al. 2021). This rapid growth has been driven by the accumulation of fuels due to a legacy of fire suppression over the last century, and a more recent increase in fuel aridity such as that observed in the western U.S., a trend which is expected to continue as the climate warms (Burke et al. 2021). Potential fire probabilities across CONUS are increasing with rising temperatures as the primary mechanism for the projected increases (Gao et al. 2021). Aside existing high-risk areas, this will also affect regions not currently associated with frequently occurring wildfires. For example, Rocky Mountain subalpine forests are now burning more than any time in recent millennia (Higuera et al. 2021). Further, forest fires in mountainous ecoregions of the western U.S. increasingly occur at higher elevations, scorching territories previously too wet to burn (Alizadeh et al. 2021). In addition, forests in the western U.S. suffer from increasing stress from insects, heat and droughts due to regional warming (van Mantgem et al. 2009). In contrast, warming is facilitating woody vegetation growth in western Alaska in the temperature-limited arctic (Song et al. 2018). Boreal forests and Arctic ecosystems respond currently with both enhanced productivity and increased respiration (Bruhwiler et al. 2021). However, climate change is predicted to alter disturbance regimes in U.S. forest ecosystems (Vose et al. 2012), and potentially affect SOC stocks. Wildfires, insect infestations, pulses of erosion and flooding, and drought-induced tree mortality are all expected to increase during the twenty-first century. For example, hydrological modeling and tree-ring reconstructions of summer soil moisture indicate that a megadrought is emerging in southwestern U.S. forests (Williams et al. 2020a).

Climate change affects important abiotic (fire, drought, wind, snow and ice) and biotic (insects and pathogens) disturbance agents in forests. Future changes in disturbance are likely to be most pronounced in coniferous forests and the boreal biome (Seidl et al. 2017). Climate change effects may likely reduce C sequestration in some forest areas in the U.S. but may also increase it in other areas (Vose et al.

2012). Regional effects of climate change include: (i) widespread changes in the distribution and abundance of dominant forest species in the Northwest; (ii) widespread dieback of pinyon pine (*Pinus edulis* Engelm.), lower forest productivity at most lower elevation locations, and lower C storage in most forest ecosystems in the Southwest; (iii) reduced tree distribution in the Great Plains; (iv) changes in abundance and extent of tree species including oak (*Quercus* spp.) decline and more common invasive species in the Midwest; (v) major reduction of spruce-fir (Picea spp. – Abies spp.) forest, moderate reduction of maple-birch beech (Acer spp. – Betula spp. Fagus spp.) forest, and expansion of oak-dominated forest in the Northeast; and (vi) reduction in the potential for C sequestration in the Southeast. However, productivity effects related to climate change and their influence on C stocks are generally unknown. Thinning and fuel treatment to reduce the vulnerability of forests to disturbance regimes and stressors may be associated with short-term changes in C stocks to enhance long-term sequestration of C (Vose et al. 2012).

Climate change mitigation through forest C management generally focuses on: (i) land use change to increase forest area (afforestation) and avoid deforestation; (ii) C management in existing forests; and (iii) use of wood as biomass energy, in place of fossil fuel or in wood products for C storage and in place of other building materials (Vose et al. 2012). Verkerk et al. (2020) called for Climate-Smart Forestry to: (i) increase the total forest area and avoid deforestation, (ii) connect mitigation with adaption measures to enhance the resilience of global forest resources, and (iii) use wood for products that store C and substitute emission-intensive fossil and non-renewable products and materials.

Invasion by nonnative (exotic, introduced, alien, invasive) plant species contributes also to global change, and this may affect terrestrial C storage. Data on the impact of many plant, animal, insect, and pathogen invasive species on SOC are, however, generally scanty (Berryman et al. 2020). AM-dominant forests are invaded by exotic plants to a greater extent than ectomycorrhizal (ECM)-dominant forests (Jo et al. 2018). Further, plots with native grass species in Texas had higher total soil C stocks to 1-m (39-in) depth (64.3 kg C m^{-2} [13.2 lbs C ft^{-2}]) than did those with nonnative grass species (61.7 kg C m^{-2} [12.6 lbs C ft^{-2}]) after eight growing seasons (Wilsey et al. 2020). The differences were, however, due to carbonate and not SOC fractions, where carbonate was 0.25 kg C m^{-2} (0.05 lbs C ft^{-2}) lower to 1-m (39-in) soil depth under nonnative than native plantings. Thus, the carbonate fraction can be an active soil C pool, and should be considered in assessments of adaptation to climate and global changes (Wilsey et al. 2020).

To sum up, the fate of the SOC stock in the U.S. under climate and global changes is uncertain. Uncertainty in the projected SOC changes under global warming is particularly high, and to minimize it requires system-level accounting for simultaneous plant productivity and microbial mineralization responses to warming and other global changes (Lugato et al. 2021).

3.4 Carbon Monitoring and Accounting

The United States Department of Agriculture (USDA) Forest Service is responsible for monitoring and reporting GHG emissions and removals from forest land in the U.S. each year (Janowiak et al. 2017). The data are a contribution to the inventory of U.S. GHG emissions and sinks prepared by the United States Environmental Protection Agency (U.S. EPA 2021). Forest Inventory and Analysis (FIA) field data, and data on harvested wood products are also considered. For reporting GHG fluxes to the United Nations Framework Convention on Climate Change (UNFCCC), GHG fluxes on all managed lands, i.e., lands were direct human intervention has influenced its condition, are accounted. The C stocks and their changes on managed forest land across CONUS, Hawaii, and southeast and south-central Alaska are estimated by the Forest Service. FIA forest inventories transitioned to a nationally consistent annualized remeasurement system starting in the late 1990s, with all CONUS states included by 2011 (Hoover and Smith 2021). The inventory design and available information as well as biomass equations for most tree species have changed over time. Only limited soil data are available for Alaska (Kimsey et al. 2020).

The basis to account for C stock changes in agricultural lands is the National Resources Inventory (NRI) conducted by the USDA Natural Resources Conservation Service (NRCS). Annual crops are considered by the U.S. EPA to be ephemeral with no net GHG emission to the atmosphere (Janowiak et al. 2017). Depending on the availability of inventory data, C accounting is done by the stock difference method, the gain-loss method or a combination of both. For the stock difference method, mean annual net C emissions or removals for land subject to human activities are estimated as the ratio of the difference in C stock estimates at two points in time and the number of intervening years. For the gain-loss method, total net emissions are first estimated as the sum of activities of products of activity area estimates and emissions factor estimates for those activities. The total is then divided by the number of intervening years to estimate mean annual net emissions or sequestration (Janowiak et al. 2017).

Soil Organic Carbon Dynamics

Temporal changes in SOC stocks for the entire U.S. have not been assessed based on repeated inventories. The assessments of soil C sequestration are, therefore, currently based on incomplete soil data and/or inappropriate soil information. To improve the knowledge basis, the USDA-NRCS Soil and Plant Science Division has launched Soils2026, an ambitious initiative to provide a new inventory of soils and provisional ecological sites for all areas of the U.S. by 2026 (Thompson et al. 2020). The target soil properties will include organic C (g kg^{-1}), rock fragments (m^3 m^{-3}), bulk density of the fine earth fraction (Mg m^{-3}), and bulk density of the whole soil (Mg m^{-3}) at the six depth intervals 0–5, 5–15, 15–30, 30–60, 60–100, and 100–200 cm (0–2, 2–6, 6–12, 12–24, 24–39, and 39–79 in). Baseline SOC stocks to 1-m (39 in) depth were recently estimated across CONUS (Wills et al. 2014). Subsequently, repeated soil sampling every 5–10 years will be needed to monitor temporal changes in SOC stocks. Adjustments in sampling design may also be needed. For example, to capture the impact of future climate by 2100 on whole-profile SOC stocks in Alaska, up to 41 additional sites would be needed in addition to the previously sampled 906 sites (Vitharana et al. 2017).

Studies on temporal changes in SOC stocks are available for some U.S. states. For example, in Florida the median SOC stock in 0–20 cm (0–8 in) depth increased from 2.69 to 3.40 kg C m^{-2} (0.55–0.70 lbs C ft^{-2}) over the past four decades (Xiong et al. 2014). The SOC sequestration rate in Florida depended on land use/land cover (LULC), and was controlled by climate factors interacting with LULC. Higher temperature tended to accelerate SOC accumulation while higher precipitation reduced the SOC sequestration rate. LULC changes observed over the past four decades also favored the SOC sequestration in top soils across Florida due to the increase in the SOC-rich wetland area by 140%, and the decrease in the SOC-poor agricultural area by 20% (Xiong et al. 2014). However, temporal changes in SOC stocks should be assessed in samples taken to at least 1-m (39 in) depth, and in certain cases soil sampling even deeper may be necessary (Gross and Harrison 2019).

Unless data from repeated soil sampling and inventories are available, SOC models and terrestrial biosphere models may be used to simulate historical dynamics in SOC stocks of the U.S., and to project future changes (Campbell and Paustian 2015; Minasny et al. 2013). However, simulations of historical and future changes in SOC stocks vary widely among models (Lajtha et al. 2018; Tian et al. 2015). Similarly, SOC projections from Earth System Models (ESMs) which include SOC cycling models are also biased (Luo et al. 2016). Globally, future net soil C sequestration rates estimated based on fifteen ESMs were 0.4‰ y^{-1} (0.6 Pg C y^{-1} [0.7 billion tn C y^{-1}]; Ito et al. 2020). For some U.S. states, future temporal changes in SOC stocks have also been simulated. For example, based on a spatially explicit prediction model, Adhikari et al. (2019) simulated that soils of Wisconsin would store an additional 20.0 Mg SOC ha^{-1} (8.9 tn SOC ac^{-1}) in 0–30 cm (0–12 in) soil depth by 2050 on top of the baseline stock of 90.0 Mg SOC ha^{-1} (40.1 tn SOC ac^{-1}). The changes are depending on soil order, land use and ecological zones. However, deeper soil depths must also be included to credibly account for future changes in SOC stocks and sequestration.

The COMET-Planner is an evaluation tool based on a soil biogeochemical model designed to provide approximate GHG mitigation potentials for NRCS CA practices (Swan et al. 2015). The summarized C sequestration and GHG emission reduction values provide generalized estimates of the GHG impacts of CA practices for initial planning purposes. The estimated SOC sequestration rates are average values over 20 years. However, site-specific conditions are required for more detailed assessments of GHG dynamics on farms (Swan et al. 2015). Further, the fundamental assumptions of the COMET biogeochemical model may not reflect current understanding of the biotic feedbacks that govern SOC dynamics (Wieder et al. 2018). For example, the COMET model with a first-order structure may overestimate transfer of C into long-lived pools, and consequently overestimate SOC storage potential (He et al. 2016).

3.5 Potential Increases in Soil Organic Carbon Stocks and Sequestration

The C fixed by photosynthesis (GPP) is the main natural source of soil C inputs, and contributor to the formation and sequestration of the SOC stock. However, terrestrial biomes differ in the percentage of fixed C remaining. The net ecosystem exchange (NEE) is a measure of the net ecosystem CO_2 exchange as the difference between GPP and total ecosystem respiration (TER; Ballantyne et al. 2021). The C exchange efficiency (CEE = NEE/GPP) is useful for comparing relative C fluxes across terrestrial biomes. At the scale of the continental U.S., 5% of the C fixed annually through photosynthesis remains in terrestrial biomes. Grasslands have very low CEE (~2%), whereas evergreen needleleaf forests and deciduous broadleaf forests appear to have quite high CEE values (~31% and ~24%, respectively). Croplands of the continental U.S. have an overall CEE of 23%. While CEE values imply that U.S. forests are strong C sinks, measurements suggest that much of this apparent forest C uptake is lost, indicating that the needleleaf and deciduous broadleaf forests may be acting more like "C sieves". Overall, identifying the biomes in which C is accumulating is difficult (Ballantyne et al. 2021), adding to the uncertainty regarding the contribution of natural processes to SOC sequestration.

In the following sections, effects of natural disturbances and human activities on SOC stocks and sequestration will be presented including agricultural land and forest stand management practices potentially resulting in increases in SOC stocks (Table 3.2). This will be done separately for the terrestrial biomes of the U.S. introduced in Chap. 1.

3.5.1 Forest Biomes

Mineral soils of U.S. forest lands sequestered C while organic forest soils lost C in the year 2019 (Table 3.1, U.S. EPA 2021). In 2001–2019, forests of CONUS were an estimated net C sink of −0.61 Pg CO_2e y^{-1} (−0.67 billion tn CO_2e y^{-1}), reflecting a balance between gross C removals (−1.39 Pg CO_2e y^{-1} [−1.53 billion tn CO_2e y^{-1}]) and gross emissions from disturbances (0.78 Pg CO_2e y^{-1} [0.86 billion tn CO_2e y^{-1}]; Harris et al. 2021). Averaged over the period 2011–2015, U.S. forests were a net sink of −0.46 Pg CO_2 y^{-1} (−0.51 billion tn CO_2 y^{-1}) while for the same period a sink of −0.65 Pg CO_2 y^{-1} (−0.72 billion tn CO_2 y^{-1}) was reported to UNFCCC as part of the national GHG inventory (NGHGI; Tubiello et al. 2021). In total, forest land, harvested wood products (HWPs), woodlands, and urban trees within the land sector collectively represent the largest net C sink in the U.S. (U.S. EPA 2021). Forest management is aimed at meeting multiple goals with that to increase C storage being one of it (Ryan et al. 2010). Carbon is a relatively new consideration in forest management, and in some situations management for forest C storage may be at odds with: (i) wildlife management activities; (ii) the provisioning of water; (iii)

Table 3.2 Soil and land-use management practices for maintaining and/or increasing soil organic carbon stocks in different biomes of the United States

Biome	Practice	References
Boreal forest/ Taiga	Rewetting forested peatlands	Nave et al. (2019a) and Ping et al. (2010)
	Nitrogen fertilization	Mayer et al. (2020)
	Reforestation of (i) stand-replacing windthrows, (ii) Picea stands with Populus or fire-adapted Pinus, and (iii) of burned slopes	Nave et al. (2019a) and Bouchard et al. (2009)
	Introducing nitrogen-fixing tree species, i.e., alder (*Alnus* spp.), black locust (*Robinia pseudoacacia* L.)	Vesterdal et al. (2013)
	Excluding ungulate herbivores	Mayer et al. (2020) and Kielland and Bryant (1998)
Temperate coniferous forest	Ignition of prescribed fires according to the historical "natural" fire return frequency	Page-Dumroese et al. (2003)
	Replacing intensive harvesting with retaining harvest residues, and reducing the extent of harvest area on a landscape level	Nave et al. (2019a)
	Managing rotation period lengths	Jandl et al. (2007)
	Replacing natural regeneration after harvest or disturbance with deliberate reforestation	Nave et al. (2019b)
	Introducing/favoring of tree species that are faster-growing, and/or more tolerant of heat or drought	Nave et al. (2019a)
	Nitrogen fertilization	Schulte-Uebing and de Vries (2018) and Johnson and Curtis (2001)
Temperate broadleaf and mixed forest	Increasing the frequency of prescribed fires and the abundance of hardwood species	Safford and Vallejo (2019)
	Retaining residues, fertilization or reforestation of wind-damaged forests	Nave et al. (2019a)
	Managing rotation period lengths	Jandl et al. (2007)
	Introduction/favoring tree species (i) accruing forest floor and mineral soil carbon, (ii) fast-growing, (iii) with high heat or drought tolerance, (iv) fixing nitrogen, and (v) associated with arbuscular mycorrhizal fungi	Keller et al. (2021), Nave et al. (2019a), Craig et al. (2018), Vesterdal et al. (2013), Wang and Huang (2020), and Hoover (2003)
Tropical forest	In hurricane-damaged forests active reforestation, facilitation of nitrogen-fixer establishment, fertilization, selection of high-SOC species/genetic families for plantations, and retaining residues during salvage logging	Nave et al. (2019a)

(continued)

Table 3.2 (continued)

Biome	Practice	References
	Reforestation with more drought resistant or drought tolerant species in tropical moist forests affected by drought events	Cusack and Marín-Spiotta (2019)
	Reforestation of secondary tropical moist forests with new tree species combinations, establishment of forest plantations, facilitation of nitrogen-fixer establishment, reforestation of degraded soils with high-soil organic carbon species, deliberate reforestation of eroded slopes, and erosion control	Mayer et al. (2020), Levy-Varon et al. (2019), Nave et al. (2019a), and Silver et al. (2003)
	Reforestation of tropical dry forests with drought resistant or drought tolerant tree species, facilitation of nitrogen-fixer establishment, addition of surface organic amendments and residue retention	Nave et al. (2019a)
	Fertilization with nitrogen and/or base cations, selection of high-soil organic carbon species or genetic families in plantations of tropical moist forests	Nave et al. (2019a)
Temperate grasslands, savannas, and shrublands	Decrease in livestock numbers, improved grazing management, fertilization (nitrogen, phosphorus, potassium and micronutrients), irrigation, increasing species diversity, and sowing legumes and improved grass species	Chang et al. (2021), Godde et al. (2020), Ochoa-Huesco et al. (2020), Conant et al. (2017), Lange et al. (2015), and Swan et al. (2015)
	Limiting post-fire erosion, suppressing severe or frequent fires	Soong and Cotrufo (2015)
	Exogenous organic matter addition	Bossio et al. (2020) and Sarkar et al. (2020)
Tundra	–[a]	Goldstein et al. (2020)
Terrestrial wetlands	Control of water levels under production agriculture	Kolka et al. (2018)
	Management of low-severity surface burns in forested peatlands and wetlands	Flanagan et al. (2020)
Desert and xeric shrublands	Introducing alfalfa (*Medicago sativa* L.)	McCarthy et al. (2018)
	Managing livestock grazing intensity	Sanderson et al. (2020) and Abdalla et al. (2018)
Croplands	Maintaining permanent cropland cover with vegetation (i.e., elimination of summer fallow, use of perennials and cover crops), protecting the soil from erosion (i.e., reduced tillage or no-till, maintaining residue cover), diverse crop rotations, improved nutrient and water management	Ledo et al. (2020), Nunes et al. (2020c), Page et al. (2020), Sainju (2020), Zhang et al. (2020a), Hristov et al. (2018), and Swan et al. (2015)

(continued)

Table 3.2 (continued)

Biome	Practice	References
	Removing highly erodible cropland from crop production, converting croplands too marginal for crop production, and cropland degraded by long-term cropping into grasslands	Sperow (2020a) and Follett et al. (2001)
	Irrigation in arid and semiarid regions	Denef et al. (2008)
Urban	–[b]	Lorenz and Shaw (2017) and Scharenbroch et al. (2017)

[a]None proven
[b]Data scanty

reduction of risks associated with undesired wildfire, insect and disease outbreaks, and invasive species; and (iv) climate change adaptation and mitigation (Janowiak et al. 2017). Climate effects of U.S. forests depend also on management-induced alterations in surface albedo, evapotranspiration and emissions of biogenic volatile organic compounds (Williams et al. 2021).

Practices that can be used to effectively manage forest C stocks include: (i) avoiding deforestation, (ii) thinning, (iii) protecting old-growth forests and other forests containing high C stocks, (iv) intensive management, and (v) forest products (Vose et al. 2012). Annual C losses from forest clearcut and thinning in CONUS amounted to 118 Tg C y^{-1} (130 million tn C y^{-1}) between 1971 and 2015 (Liu et al. 2020c). Otherwise, the annual C sequestration rate has been basically the same in recent years due to a combination of increased C gain from net primary productivity (NPP) and increased C loss from heterotrophic respiration, land use and cover change, and fire disturbances (Liu et al. 2020a). In the future, NPP may increase by elevated CO_2, but the response may diminish over time (Norby and Zak 2011). At the whole-tree level, CO_2 fertilization causes consistent biomass increments in young tree seedlings only, whereas mature trees show a variable response (Lauriks et al. 2021). Overall, Wear and Coulston (2015) projected a gradual decline in the forest C sink in CONUS over the next 20 years albeit with regional differences.

Effects of Common Forest Management Practices

The forest SOC stocks and sequestration may be affected by practices including harvesting, thinning, fertilization, liming, drainage, irrigation, tree species selection, and control of understory vegetation and natural disturbances (Binkley and Fisher 2020; Jandl et al. 2007). Mayer et al. (2020) summarized current knowledge about the effects of 13 common forest management practices on forest SOC stocks. For example, afforestation of former croplands generally increases SOC stocks, whereas on former grasslands and peatlands, SOC stocks are unchanged or even reduced following afforestation. Afforestation with evergreen broadleaf species may generally support the formation of SOC (Hou et al. 2020). Cook-Patton et al. (2020a) identified opportunity classes for restoration of forest cover across

CONUS. An area of up to 516,000 km^2 (199,000 mi^2) could be restored to forest cover which could potentially capture 0.31 Pg CO_2 y^{-1} (0.35 billion tn CO_2 y^{-1}). Pastures hold the most C capture potential at lowest cost. Non-stocked forest patches and federal lands also hold substantial low-cost opportunity, whereas watersheds and urban areas hold high potential to capture C, but are more costly (Cook-Patton et al. 2020a).

Few management studies show gains in forest SOC stocks, and reducing the vulnerability to SOC loss may be a more conservative approach to SOC stewardship in forests (Nave et al. 2019a). For example, stem-only and whole-tree harvesting may have negligible impacts on SOC stocks (Johnson and Curtis 2001; Nave et al. 2010; Thiffault et al. 2011). However, saw log harvesting can increase the SOC by as much as 18% while whole tree harvesting may decrease it by 6%. In contrast, much less is known about the impact of thinning on SOC stocks (Jurgensen et al. 2012; Nave et al. 2010).

The conversion of primary forests to secondary forests may reduce SOC stocks, particularly if the soil is converted to an agricultural land-use prior to reforestation (Mayer et al. 2020). Primary forests are naturally regenerated forests of native species where there are no clearly visible indications of human activities, and the ecological processes are not significantly disturbed (Bernier et al. 2017). Reforestation may sequester 2.0 Pg C y^{-1} (2.2 billion tn C y^{-1}) to 10-cm (3.9-in) soil depth in CONUS in a century (Nave et al. 2018). Fargione et al. (2021) emphasized the need for public support for investing in reforestation activities and incentives for landowners. Reforestation in the studied scenario referred to any tree planting that causes tree cover to increase to more than 25% on lands where forests historically occurred based on modeled potential vegetation. This included lands that recently had forest cover, as well as lands that have been in a non-forest land use for an extended period. About 260,000 km^2 (100,387 mi^2) of natural and agricultural lands in CONUS would be reforested by 2040 with 30 billion trees at an estimated cost of $33 billion USD. This would require increasing the number of tree seedlings produced each year by 1.7 billion, a 2.3-fold increase over current production levels. In addition, investments would be needed to expand capacity for seed collection, seedling production, workforce development, and improvements in pre- and post-planting practices (Fargione et al. 2021).

If all understocked timberlands were fully stocked, potential live tree biomass C sequestration capacity in CONUS would increase by about 51 Tg C y^{-1} (56 million tn C y^{-1}). Understocked means the <35% of the forestland area is occupied by trees. In comparison, the annual sequestration rate in live tree biomass (above- and below-ground) was about 0.15 Pg C (0.17 billion tn C) in the year 2018. However, the forest C sink may be slowly diminishing (Domke et al. 2020). Natural forest regrowth, i.e., recovery of forest cover on cleared lands through spontaneous regrowth after cessation of previous disturbance or land use, may accumulate up to 88 Tg C y^{-1} (97 million tn C y^{-1}) in tree biomass in the U.S. over 30 years (Cook-Patton et al. 2020b).

Harvesting (i.e., clear-cut harvesting) generally results in a reduction in SOC stocks, particularly in the forest floor and upper mineral soil (Mayer et al. 2020).

Specifically, removal of residues by harvesting whole-trees and stumps negatively affects SOC stocks. Soil disturbance from site preparation also decreases SOC stocks, particularly in the organic topsoil. Harvesting forests may prevent the high litter input required to reverse POM losses and exacerbate negative climate feedbacks (Lugato et al. 2021). A shift to forest management that increases structural litter inputs to the soil would, instead, promote the formation of POM. Similarly, James et al. (2021) suggested that intensive removal of forest residues could cause persistent losses of soil C. However, improved growth of tree seedlings may potentially outweigh some of the SOC losses over a rotation cycle (Mayer et al. 2020). Modest residue removals (<80%) of biomass using whole tree harvest with forest floor retention could provide a sustainable source of biomass for bioenergy production without additional soil C losses compared to bole-only harvesting practices (James et al. 2021).

The addition of N has an overall positive effect on SOC stocks across a wide range of forest ecosystems (Mayer et al. 2020). Higher SOC stocks are also found under tree species with N-fixing associates. The SOC stocks differ under different tree species, with coniferous species accumulating more organic C in the forest floor and broadleaved species tending to store more SOC in the mineral soil (Mayer et al. 2020). There is some evidence that increased tree species diversity could positively affect SOC stocks in temperate and subtropical forests (Mayer et al. 2020). However, tree species identity, particularly N-fixing species, seems to have a stronger impact on SOC stocks than tree species diversity. For example, positive effects of plant species-mixture on forest SOC stocks were reported based on a global meta-analysis (Chen et al. 2020a). Specifically, high plant diversity in forests decreased litter C/N ratios and, thus, indirectly promoted SOC accumulation by increasing the litter quality (Zhou et al. 2019b). Also, leguminous plants with low C/N ratios may contribute to increases in forest SOC (Johnson and Curtis 2001). For example, mineral soil C stocks (median depth 10 cm [3.9 in]) were increased (9%, 10%, and 6%) under broadleaved, AM, and N-fixing trees compared to that under coniferous, ECM or non-N fixing trees. In contrast, forest floor C stocks were lower under broadleaved (44%), AM (39%), or N-fixing (28%) trees compared to that under coniferous, ECM, or non-N fixing trees (Peng et al. 2020). However, understanding of quantitative effects of tree species change on profile SOC stock in forests is limited (Mayer et al. 2020). For example, in less productive environments, promoting the abundance of locally adapted and stress tolerant tree species instead of increasing species richness may contribute to increases in the C storage capacity of forests (Madrigal-González et al. 2020). Both fertilization and N-fixing trees can cause marked overall increases in SOC (Johnson and Curtis 2001). Typically, N-fixing trees add 15–20 g (0.5–0.7 oz) of stabilized SOC for each gram (0.03 ounces) of N accumulated (Binkley and Fisher 2020). However, the effect of N-fixing trees on forest development and C storage in CONUS depends on local factors (Staccone et al. 2021). Thus, more studies in the U.S. are needed that focus on N-fixing trees, such as Alnus species (Berryman et al. 2020).

The management of stand density and thinning have small effects on forest SOC stocks (Mayer et al. 2020). In forests with high populations of ungulate herbivores, reduction in herbivory levels can increase SOC stocks. Removal of plant biomass for fodder and fuel may be related to a reduction in SOC stocks. Fire management practices such as prescribed burning also reduces SOC stocks, but less so than wildfires which are more intense (Mayer et al. 2020). However, observational evidence on how fire affects forest productivity and SOC stocks is limited (Lasslop et al. 2020). Detailed long-term fire records for CONUS are also absent (Liu and Yang 2021).

Fire plays an integral role in structure and function of forest ecosystems (Binkley and Fisher 2020). Fire generates pyrogenic OM which is a more recalcitrant soil C input than other plant-derived inputs. Specifically, the pyrogenic OM in soil has a long turnover time of centuries to millennia (Kuzyakov et al. 2009). Pyrogenic OM may also interact with native SOC but dynamics and mobility of pyrogenic OM in soil are largely unknown (Schiedung et al. 2020). Nevertheless, Li et al. (2021) suggested that that fire-induced pyrogenic OM production should be accounted for in the C budget. Overall, forest fires may have no effect on SOC stocks (Lasslop et al. 2020). However, there can be an effect of time since fire as, for example, SOC stocks may increase compared to unburned control 10 years after fire occurrence (Johnson and Curtis 2001) or be similar (Li et al. 2021). Otherwise, SOC stocks can be lower following prescribed fire or supervised burns but higher compared to unburned sites following wildfire.

Management of stand-replacing disturbances may be particularly promising to maintain and/or increase SOC stocks as those cause C losses from forest ecosystems (Amiro et al. 2010). However, challenges to forest management arise in mountainous ecological zones with steep altitudinal gradients and island effects (Nave et al. 2019a). Wetland forest SOC stocks are highly vulnerable to management related changes in wetland hydrology or vegetation conversions.

Globally, there is ample evidence about the effects of management on SOC stocks in the organic layers of the forest floor, but much less information about effects of management on mineral soil SOC stocks (Jandl et al. 2007). The forest SOC stock can be enhanced by increasing forest productivity and thereby increasing the C input to the soil. Minimizing the disturbances in the stand structure and soil reduces the risk of unintended SOC losses. The establishment of mixed species forests increases the stability of the forest, and can avoid high rates of SOC decomposition. Shallow rooting coniferous tree species tend to accumulate forest floor C compared to hardwood tree species but the extent of the effect of tree species on SOC stocks across site types is uncertain (Jandl et al. 2007). Management decisions on enhancing SOC in forest soils can be supported by ecosystem-level forest models that incorporate the most important ecological processes influencing tree growth and soil development (Blanco 2018). How management of existing forests and not afforestation of non-forest land may alter SOC stocks and sequestration in different forest biomes of the U.S. will be discussed in the following sections. Overall, SOC stocks of U.S. forests appear not to be particularly sensitive to land cover type, and to typical land use decisions (Nave et al. 2019a).

Boreal Forest/Taiga Biome

Similar to C stocks in Canada's boreal forests, climate change may be the single biggest threat to boreal C stocks in the U.S. (Kurz et al. 2013). There is some knowledge on the C balance of managed boreal forests in Alaska but vast areas are unmanaged leaving gaps in monitoring, observation and quantification of boreal forest C dynamics (Safford and Vallejo 2019). The boreal forests of Alaska, like those of Canada, have not been subjected to extensive land-use change, and their C dynamics are largely driven by natural disturbances (McGuire et al. 2006).

Natural disturbance regimes influence biogeochemical cycling and landscape composition in boreal forests (Thiffault 2019). From 1982 to 2016, forests in eastern Alaska, for example, exhibited large patches of tree canopy loss and short vegetation gain as a result of persistent disturbances from wildfires and subsequent recovery of natural vegetation (Song et al. 2018). In northwestern North American boreal forests, disturbances suppressed the aboveground C sink from 1984 to 2014 as fire-and harvest-related C losses were not fully compensated (Wang et al. 2021a). This reduction, in addition to belowground changes such as soil warming, permafrost thaw and high-latitude peatland fires may also influence the boreal forest SOC stock (Wang et al. 2021a).

Some studies on the effects of natural disturbances on SOC stocks in boreal forests are available. Fire is the dominant source of boreal forest SOC vulnerability which is enhanced by expanding mineralization seasons, accelerating permafrost thaw, and insect outbreak – fire interactions (Nave et al. 2019a). Some boreal fires exhibit 'overwintering' behaviour, in which they smoulder through the non-fire season and flare up in the subsequent spring. This process is favored by the deep organic soils and accelerated warming, with 38% of the annual burned area in Alaska in 2008 originating from a single overwintering fire (Scholten et al. 2021). Importantly, slow forest growth and large O-horizons mean C losses can be very large and slow to recover (Nave et al. 2019a). Time since fire is the main direct driver of SOC stock increases in the boreal forest mineral soil (Andrieux et al. 2018).

In the most productive and accessible portion of the boreal forest biome, forestry is the most widespread human management activity with effects on the C cycle (Schuur et al. 2018). However, management effects on boreal forest biome SOC stocks are poorly quantified, and mechanisms behind accumulation of SOC in boreal forest soils, in particular, are not sufficiently understood (Douglas et al. 2014). The boreal forest biome is also subject to exploration, development, and extraction of other resources (mining, oil and gas, flooding for hydroelectric projects etc.; Gauthier et al. 2015) which all may affect its SOC stocks.

Natural disturbances

Fire is the best characterized and greatest ecologically important disturbance in the boreal biome affecting stand structure, species composition and C cycling through patterns of mortality and regeneration (Schuur et al. 2018). Most boreal forest fires occur in mature black spruce (*Picea mariana* Mill BSP) stands but younger and more deciduous-dominated or mixed stands are now also tending to burn (Rogers et al. 2020). The combustion of organic soils dominates boreal fire C emissions (Walker et al. 2020). Boreal forest fires are associated with long periods of forest regrowth to re-sequester lost C. For example, chronosequence studies of stand-replacing fires in boreal forests of Alaska indicate that those become net C sinks after about 10 years (Amiro et al. 2010). In some boreal landscapes recently affected by fire, soil respiration often exceeds GPP, resulting in a net annual source of CO_2 to the atmosphere (Watts et al. 2021). Nevertheless, fire in boreal forests is the dominant source of SOC vulnerability, expanding mineralization seasons and accelerating permafrost thaw (Nave et al. 2019a). Fire can also result in transition to a less productive boreal forest. The degree to which fire influences long-term boreal biome dynamics depends on fire severity, return interval, and burn depth. For example, the C stocks of the organic layer may need on average ≥100 years for full recovery from fire (Palviainen et al. 2020). Otherwise, in Alaskan boreal forests shifts in dominant plant species catalyzed by severe fire may compensate for greater combustion of SOC over decadal time scales (Mack et al. 2021). Specifically, severe burning of organic soils may shift tree dominance from slow-growing black spruce to fast-growing deciduous broadleaf trees, resulting in a net increase in C storage by a factor of 5 over the disturbance cycle (Mack et al. 2021).

Wildfires impose natural limits on the rotation period (Jandl et al. 2007). Fire has a large direct and immediate impact on C cycling as the combustion process causes the release of trace gases such as CO_2, methane (CH_4), and carbon monoxide (CO), and generates recalcitrant pyrogenic OM or black carbon (charcoal), primarily through the combustion of SOM (Harden et al. 2000). Management strategies to reduce fire-related SOC losses may include decreasing fire extent/severity by incorporating young stands into the landscape mosaic (Nave et al. 2019a). However, effects of stand age on SOC loss vary. For example, intermediate-aged burned black spruce stands (37–52 years old) in interior Alaska had much less remaining C in their organic soils compared to burned mature stands (70–120 years old; Hoy et al. 2016). In addition, drainage, peat moss (*Sphagnum* spp.) and permafrost influence spatial variation in fire C losses. The SOC stocks on boreal mountain forests can be increased by reforestation of burned slopes (Nave et al. 2019a).

Current-day fire C emissions from boreal ecosystems are roughly balanced out by post-fire legacy C sinks (Yue et al. 2016). A net accumulation of C can occur over multiple fire events when the soil beneath the burned layer escapes combustion (Walker et al. 2019). Boreal forest burning in Alaska averaged 0.025 million km^2 (0.010 million mi^2) per year during the last half century (Schuur et al. 2018). Based on a modeling experiment, changing fire management planning options would decrease the projected cumulative area burned by 1.5–4.4% by 2100 with proportional effects on C emissions. Otherwise, frequently burned boreal needleleaf forests experienced no changes in surface SOC stocks (Pellegrini et al. 2018). Thus, fire management may not alter SOC stocks and sequestration in the boreal forest biome in the long-term.

Boreal forest fires continue to increase in size, severity and intensity, and this will potentially threaten the C sink of boreal forest soils in the future (Walker et al. 2019). Climate change-induced modification of fuels needs to be considered for accurately predicting future C emissions from boreal wildfires (Walker et al. 2020).

Although climate change likely shortens the fire return interval, increased potential evapotranspiration resulting from temperature increases may speed up C accumulation and the recovery rate of C stocks in boreal forests (Palviainen et al. 2020). The balance between these two processes will ultimately determine the SOC stocks of the boreal forest. As the boreal region continues to warm, soil respiration may increasingly offset GPP with unknown consequences for SOC stocks (Watts et al. 2021). More studies in boreal forests are needed to test the generality in the response and application over longer fire-return intervals, also for severe crown fires that can consume the forest floor. Further, studies of gradients in long-term fire frequencies are lacking and do not always examine changes in mineral SOC stocks (Pellegrini et al. 2018). The deep SOC stock in boreal forest soils may be sensitive to multi-decadal changes in fire frequencies similar to that in temperate savanna soils (Pellegrini et al. 2021).

Insect and pathogen outbreaks in boreal forests can also kill or damage trees, and potentially decrease forest floor C and SOC stocks or have no effects but long-term impacts on SOC stocks and sequestration are uncertain (Hicke et al. 2012). Specifically, there is no clear evidence on how insect herbivory affects forest C-storage on longer time scales (Kristensen et al. 2020). Nevertheless, SOC losses from boreal forests can probably be reduced by more aggressive management of insect outbreaks but those have been proven difficult to contain (Kurz et al. 2008).

Windthrow is a natural phenomenon by which strong winds shear or completely uproot trees (Overby et al. 2003). Where fire is rare, it is the predominant natural disturbance in boreal forests (Bormann et al. 1995). Thus, boreal forests that have evolved with windthrow may have minimally altered forest floor and mineral SOC stocks over long time periods (Liechty et al. 1997). The potential to increase SOC stocks and enhance sequestration by managing windthrow effects may, therefore, be small except for stand-replacing windthrows defined as a continuous area of at least 5 ha (12 ac) where post-windthrow canopy cover is less than 25% (Bouchard et al. 2009).

Stand management

Most boreal forest area in Alaska is under federal management, and much is pristine or near-pristine forest with little management or restoration work other than local wildlife habitat improvement (Safford and Vallejo 2019). Losses from the SOC stock may be modest when primary boreal forests are converted directly to managed forests (Kishchuk et al. 2016). The connectivity between relatively pristine forests across managed boreal forests that separate them is among the restoration aims to reduce genetic isolation and permit species migration in the boreal forest biome in Alaska. Heterogenous boreal forest landscapes such as mixtures of open and closed habitats, and mixtures of coniferous and deciduous species should be favored to enhance resilience to large, synchronous disturbances (Safford and Vallejo 2019), thereby reducing effects on SOC sequestration and stocks.

Draining boreal forested peatlands can increase short-term C emissions but also long-term SOC storage (Minkkinen et al. 2002; Toberman et al. 2010). In contrast, Ping et al. (2010) found that SOC in black spruce forests in Alaska tended to increase as water drainage decreased. Maintaining saturated conditions in Histosols under boreal forest cover are critical to protect SOC (Nave et al. 2019a). Drying cycles are common during summer months in boreal forests and likely play a significant role in affecting the net C efflux from soil (DeLuca and Boisvenue 2012). Thus, creation of saturated soil conditions results in slowed decomposition rates. Felling residual wood may drive conversion over time to peatland (paludification), and result in increases in SOC stocks (Nave et al. 2019a). However, while peat moss-dominated peatlands accumulate SOC by water table management and rewetting (Lucchese et al. 2010), effects on SOC stocks and sequestration in rewetted boreal forested peatlands are uncertain. While the long-term cooling effect of increased SOC sequestration in rewetted peatlands exceeds the short-term warming impact of simultaneous increases in CH_4 emissions, it is unclear how peatland restoration modifies regional climates by biophysical impacts (Helbig et al. 2020).

Following harvest, there can be an initial release of C from boreal forests depending on silvicultural practice, rotation period length and harvest-related damage (Bradshaw et al. 2009). Otherwise, losses of SOC can be reduced by lighter harvest removals (thin *vs.* clearcut, retaining residues and stumps; Nave et al. 2019a). Forest floor SOC stocks may be particularly reduced after harvesting due to low leaf and wood litterfall, and increased OM decomposition due to soil disturbance and changes in microclimatic conditions (Achat et al. 2015). However, inputs of dead root C and DOC into the mineral soil may compensate for forest floor C losses, and may result in no overall changes in boreal forest SOC stocks in regenerating forests after timber harvest (Goldstein et al. 2020).

Intensifying biomass harvest by exporting the logging residues may result in losses from the boreal forest SOC stock (Achat et al. 2015). For example, whole-tree harvesting generally leads to either no significant difference or a small reduction in SOC stocks compared with stem-only harvesting, most often in the organic layer (Clarke et al. 2015). In the long term, changes in SOC stocks of boreal forests in the U.S. are unlikely to be detectable between successive harvests (Dean et al. 2017; Goldstein et al. 2020). However, when a primary boreal forest is logged and then subjected to harvesting cycles, a long-term (i.e., ≥300 years) decrease in SOC stocks may be observed (Dean et al. 2017). Similarly, logging scars – the vestigial remains of roads, landings and turnoffs meant to accommodate heavy machinery – may hinder the natural renewal of boreal forests (Wildlands League 2019). Logging scars may also contribute to lower SOC stocks. However, the impact of management interventions such as elongation of the harvest rotation period, enhancing regeneration, or increasing stocking densities on SOC stocks of boreal forests are small compared to those of fire (Bradshaw et al. 2009).

Fertilization of boreal forests improves tree growth/litter inputs and inhibits decomposition, which increases O horizon depths and, thus, SOC stocks (Nave et al. 2019a). Specifically, greater accumulations of humus are often noted following N fertilization of boreal forests (Mayer et al. 2020). Addition of N also enhances aboveground woody biomass in boreal forests (Schulte-Uebing and de Vries 2018).

Thinning of managed boreal forests may enhance wood production, regulate stand structure, decrease severe wildfire events, and increase forest resilience to environmental disturbances (Zhang et al. 2018). Forest thinning influences SOC stocks in both direct (more litterfall, woody debris, dead root biomass) and indirect (stimulation of litterfall decomposition by changes in microclimate) ways. Thinning interventions potentially reduce forest floor C stocks by temporal increases in decomposition rate and reduced litterfall (Jandl et al. 2007). However, the input of thinning residues may compensate for C losses resulting in no overall changes in boreal forest SOC sequestration and stocks.

Reforestation of managed boreal forests following harvesting or natural disturbances may enhance SOC stocks but knowledge is limited (Nave et al. 2019b). Historic rates of deforestation and afforestation in Alaska are low comparable to the situation in Canada's boreal zone (Kurz et al. 2013). Reforestation could, however, also be used to speed the post-disturbance recovery of unmanaged boreal forests (Gauthier et al. 2015). With regards to the projected warming and increased precipitation, seed stock, genotypes and species from warmer, more southern latitudes should be introduced (Safford and Vallejo 2019). The selection of certain boreal tree species for reforestation may particularly affect SOC stocks. For example, boreal forest floor C stocks were lower for the genera Betula and Populus than for Picea (Vesterdal et al. 2013). Pinus had intermediate forest floor C stocks but similar to those of Picea. Forest floor C stock for the deciduous conifer Larix did not consistently differ from either conifers or broadleaves. Boreal forest floor C stock increased by 20–70% from Betula or Populus compared to Picea. Transition from Picea to fast growing Populus with high SOC stocks or to fire-adapted Pinus with fast SOC recovery are among options for increasing boreal forest SOC stocks (Nave et al. 2019a). However, site-related factors may mask the effects of tree species on forest floor C stocks, and it is difficult to identify a pure tree species effect on boreal SOC stocks (Vesterdal et al. 2013). In theory, converting all aspen-Betula to Abies-Picea forest-types in Alaska may increase forest floor C stocks by 0.06–0.19 Pg C (0.06–0.21 billion tn C), 0.15–0.53 Pg C (0.17–0.58 billion tn C) and 0.16–0.57 Pg C (0.19–0.63 billion tn C) in 0–30 cm (0–12 in), 0–100 cm (0–39 in) and 0–150 cm (0–59 in) depth, respectively. This was calculated based on area and SOC density data reported in Johnson and Kern (2003). However, achieving this increase is highly unlikely as boreal forest species composition is not managed for storing C but for increasing the productivity of trees for timber.

Higher boreal forest mineral soil SOC stocks with proportional differences within the range 1.2–1.8 were reported for conifers (Picea, Pinus, Larix) compared to that of broadleaves (Betula, Populus; Vesterdal et al. 2013). Based on data reported in Johnson and Kern (2003), converting the total boreal broadleaf area to conifers in the U.S. corresponds to a theoretical but unlikely achievable potential to increase mineral SOC stocks by 0.21–0.85 Pg C (0.23–0.94 billion tn C), 0.47–1.87 Pg C (0.51–2.06 billion tn C) and 0.63–2.51 Pg C (0.69–2.77 billion tn C) in 0–30 cm (0–12 in), 0–100 cm (0–39 in) and 0–150 cm (0–59 in) depth, respectively. The effect of species on mineral SOC may be stronger or more similar in direction to that of forest floor C stocks, but conclusions are hampered by paucity of data (Vesterdal et al. 2013). N-fixing tree species such as alder (*Alnus* spp.) and black locust (*Robinia pseudoacacia* L.) may be associated with higher SOC stocks in boreal forests but again data are scanty. Similarly, it is unclear whether increasing tree species diversity will affect boreal forest SOC stocks, and whether site-related factors may mask the effects of tree species on mineral soil SOC stocks (Vesterdal et al. 2013). Thus, there may be a potential to enhance SOC sequestration and increase SOC stocks in managed parts of boreal forests of the U.S. by planting selected tree species but more data are needed.

Excluding ungulate herbivores may result in increases in SOC stocks of boreal forests (Mayer et al. 2020). For example, moose [*Alces alces* (Linnaeus, 1758)] exclusion in boreal Alnus and Salix stands of the Alaskan taiga increased SOC stocks by 54% (Kielland and Bryant 1998). However, only a few studies have investigated effects of herbivores on boreal forest SOC stocks (Mayer et al. 2020). Increased human activities contribute to the expansion and growth of large herbivores in many boral forest regions (Leroux et al. 2020). The associated spatial heterogeneity in herbivory has the potential to shift boreal forests from C sink to C source. Herbivore impacts on leaf litter, in particular, will have indirect effects on the soil organic horizon (Leroux et al. 2020).

To sum up, it is not possible to fully control natural disturbances because of the scale and remoteness of the U.S. boreal forests and the characteristics of the disturbance agents involved. Better control of natural disturbances may be economically impossible, especially with regards to future climate conditions (Gauthier et al. 2015). Further, the effects of silvicultural management on SOC stocks in boreal forests may be small but those effects are often not studied (Zhu and McGuire 2016). There is insufficient knowledge to guide land managers in specific situations to change boreal forest management practices to increase C sequestration (Bhatti et al. 2012). Species selection for reforestation with SOC-accruing trees may be a promising management strategy. Opportunities for increasing SOC stocks include felling residual wood, species transition from Picea to aspen or fire-adapted Pinus, fertilization, and reforestation of burned slopes (Nave et al. 2019a).

Temperate Coniferous Forest Biome

Nearly all temperate coniferous forests in CONUS are considered managed lands (Domke et al. 2018). Temperate coniferous forests are managed for timber production, driven by an ever-increasing demand for paper, construction materials and furniture, and resulting from relatively rapid growth among overstory dominant tree species (Gilliam 2016). The lengthening of harvest cycles on private lands, and restricting harvest on public lands may generally increase the net ecosystem carbon balance (NECB) of temperate forests (Law et al. 2018), and enhance SOC sequestration. Aside harvest practices, other key SOC vulnerabilities are associated with wildfire, and its importance increases as the climate gets drier especially in seasonally dry temperate coniferous forests (Nave et al. 2019a).

Natural disturbances

The western U.S. is among the global regions with the largest increases in wildfire (Rogers et al. 2020). Climate, lightning ignitions and insect outbreaks potentially influence burned area and fire severity. From 1992 to 2012, lightning-caused fires burned in total 0.06 million km^2 (0.02 million mi^2) in the ecoregions Marine

West Coast Forests and Northwest Forested Mountains, both predominantly covered by temperate coniferous forests (Balch et al. 2017). Reburning is also increasing across the U.S. West (Buma et al. 2020). In the Pacific Northwest, frequency and extent of forest fires are projected to increase in the future (Halofsky et al. 2020). Changing fire together with human disturbance regimes, and climate may also result in fire-driven forest conversions. Specifically, conifer-dominated forests in the western U.S. may be replaced by vegetation dominated by genera of re-sprouting broadleaf trees (e.g., aspen, birch), shrubs (e.g., oak, ceanothus [*Ceanothus* L.]), and herbaceous communities (Coop et al. 2020). Fire affects forest C storage, and it may take decades after fire occurrence until a temperate coniferous forest is again a net C sink, especially under moisture limited conditions (Amiro et al. 2010). The importance of fire for SOC vulnerability also increases as climate dries, particularly in seasonally dry forests (Nave et al. 2019a).

Similar to the effects of harvesting, fire reduces forest floor SOC stocks in temperate coniferous forests, i.e., on average by 68% (Nave et al. 2011). However, fire-induced forest floor C losses are not permanent, and may recover to pre-fire levels within 100–130 years. In contrast, mineral SOC stocks in the northwest U.S. dominated by coniferous forests declined on average by 19% after fire (Nave et al. 2011). For example, Bormann et al. (2008) observed a strong reduction in SOC by an intense fire but other direct evidence of fire effects on SOC stocks is limited. Low-intensity frequent fires in coniferous forests may transform SOM in ways that may offset ecosystem C losses (Pellegrini et al. 2021). While these fires reduced forest floor C by 77%, fire also reduced C in mineral soils that was free from minerals by 41% but not C associated with minerals. Specifically, fire transformed the properties of remaining mineral SOM by increasing the proportion of pyrogenic OM (from 3.2% to 7.5%) and associated with minerals (from 46% to 58%), suggesting that the SOC remaining after low-intensity frequent fires in coniferous forests is more resistant to decomposition (Pellegrini et al. 2021).

The SOC stocks in temperate coniferous forests may be maintained or increased by prescribed fires ignited according to the historical "natural" fire return frequency as prescribed fires may cause a smaller loss in SOC stocks compared to that of wildfires (Page-Dumroese et al. 2003). For example, an average 19% increase in SOC stocks in 0–30 cm (0–12 in), 0–100 cm (0–39 in), and 0–150 cm (0–59 in) soil depth may result in maximum increases of 0.15, 0.27 and 0.30 Pg C (0.16, 0.29 and 0.33 billion tn C) for Douglas-fir (*Pseudotsuga menziesii*; Gilliam 2016) forest type, respectively. Potential increases for the same depths amount to 0.03, 0.05 and 0.06 Pg C (0.03, 0.06 and 0.06 billion tn C) for Hemlock-Sitka spruce (*Tsuga* spp. – *Picea sitchensis* [Bong.] Carr.) forest type; 0.08, 0.14 and 0.16 Pg C (0.09, 0.16 and 0.18 billion tn C) for Lodgepole pine (*Pinus contorta*) forest type; 0.17, 0.30 and 0.34 Pg C (0.19, 0.33 and 0.37 billion tn C) for Ponderosa pine (*Pinus ponderosa* Douglas ex C.Lawson) forest type; and 0.09, 0.19 and 0.24 Pg C (0.10, 0.21 and 0.27 billion tn C) for Western spruce-fir forest type. These potential increases in SOC stocks by fire management according to forest types are based on data reported in Johnson and Kern (2003). The geographic variation in fire effects on mineral

SOC stocks in temperate coniferous forests also indicate the need for region-specific fire management plans (Nave et al. 2011).

Insects and pathogens affect up to 0.2 million km^2 (0.08 million mi^2) of CONUS forestland annually (Krist et al. 2007). The largest C cycle impacts result from insects that kill trees such as the widely reported outbreak of bark beetles in temperate coniferous forests of the western U.S. that have impacted 0.15 million km^2 (0.06 million mi^2) of forestland from 1997 to 2014, and another 0.15 million km^2 (0.06 million mi^2) in the south since 1980 (Williams et al. 2016). About 5–15 Tg C y^{-1} (6–17 million tn C y^{-1}) live biomass may have been killed by bark beetle attacks across beetle-affected areas in western US forests from 2000 to 2009 (Ghimire et al. 2015). This may alter the transformation of dead material into SOC, and create a transient increase in downed wood on the forest floor (Williams et al. 2016). However, rapid growth of surviving individuals could compensate some of the live biomass reduction which amounted to a reduction in net ecosystem productivity (NEP) of 6.1–9.3 Tg C y^{-1} (6.7–10.3 million tn C y^{-1}) in 2009 (Ghimire et al. 2015). Overall, there is a positive net change in forest C stocks across CONUS, indicating forest resilience and apparently large responses to growth enhancements such as fertilization by CO_2 and N (Williams et al. 2016). Data on forest floor and SOC impacts of insect and pathogen disturbances over landscape scales are, however, variable (Hicke et al. 2012; Overby et al. 2003; Prăvălie 2018). Thus, the potential to manage SOC stocks of temperate coniferous forests in the U.S. by management of insects and pathogens is unknown.

Storm and windthrow damage may increase the input of coarse woody debris to the forest floor but reduce soil C input through the loss of photosynthetically active tissues (Jandl et al. 2007). However, SOC stocks of temperate coniferous forests that have evolved with storms and windthrow may not be altered if wind disturbance occurs within the range of natural variability (Liechty et al. 1997). Management options to manage SOC stocks and enhance sequestration by managing storm and windthrow damage in the U.S. appear, therefore, also limited.

Stand management

Forest management practices typically involve a combination of harvesting, stand regeneration/reforestation, and stand tending (Lajtha et al. 2018; Nave et al. 2019b). These practices affect temperate coniferous forests across all spatial scales, from the landscape to the overstory, and the herb layer. Management may prioritize practices targeting the renewal of older temperate conifer forests as those have lower resource use efficiencies than younger ones (Xu et al. 2020a, b). Aside for wood products, forest management is also for water and recreation services, with C uptake as a secondary outcome (Domke et al. 2018). In many temperate coniferous forest regions, C stocks are recovering from historical clearing, cultivation and thinning. This recovery stimulates forest C uptake from both afforestation/reforestation and C accumulation in still-maturing stands. Forest management also has: (i) altered forest species composition (e.g., with the establishment of plantations); (ii) generally accelerated C accumulation rates; and (iii) modified forest soil fertility, both through nutrient (i.e., N) gains from fertilizer application and nutrient losses from erosion caused by some harvesting practices. However, the net effect of such activities on forest C stocks and fluxes is uncertain (Domke et al. 2018).

Temperate coniferous forest SOC stocks, particularly those in the forest floor, may be affected by harvesting, site preparation and plantation management (Johnson and Curtis 2001; Nave et al. 2010). However, within about 20 years after harvest, GPP may recover with faster recovery of the C sink of U.S. temperate coniferous forests in warmer climates (Amiro et al. 2010). Usually, harvest impacts only the forest floor or O-horizon (Nave et al. 2019a). For example, the conventional harvesting of tree stems only in temperate coniferous forests results in SOC losses from the forest floor that are compensated by SOC gains in the mineral soil, resulting in no overall impacts of conventional harvesting on forest SOC stocks (Achat et al. 2015). In contrast, intensive harvesting where in addition to tree stems logging residues are removed to produce fuelwood may cause a loss of SOC in all forest soil layers. Thus, retaining harvest residues is a management option to reduce SOC losses (Nave et al. 2019a). For example, residue retention in temperate oceanic forests of the western U.S. growing on Cambisols may promote SOC accumulation. Otherwise, reducing disturbance and continued C inputs via detritus and soluble OM can maintain SOC in temperate oceanic forests growing on Podzols (Nave et al. 2019a). However, published studies on intensive harvesting effects on complete organic plus mineral soil profile SOC stocks are scanty (Achat et al. 2015). Overall, SOC stocks recover to pre-harvest levels after decades to a century or more (Lajtha et al. 2018). Nevertheless, reducing the extent of harvest area on a landscape to allow long-term C accumulation in soil and vegetation is an opportunity to manage SOC stocks (Nave et al. 2019a).

A long-term decrease in SOC stocks may occur when a primary coniferous forest is logged, and then subjected to harvesting cycles (Dean et al. 2017). The rotation period length also plays a significant role in the degree of harvest impacts on SOC. For example, 100 years post-harvest the SOC stocks in the organic layer and in 0–50 cm (0–20 in) depth in a managed northeastern red spruce (*Picea rubens* Sarg.) forest chronosequence in Nova Scotia, Canada, approached the range of the intact forest while those were only half of the intact forest after 32 years post-harvest (Diochon et al. 2009). Thus, preserving SOC stocks in successional temperate coniferous forest stands in the U.S. may be achieved by managing rotation period lengths for maximizing SOC stocks (Jandl et al. 2007). For example, assuming that white-red-jack pine (*Pinus strobus* L.-*Pinus resinosa* Ait.-*Pinus banksiana* Lamb.) forests in the U.S. are harvested before peak SOC stocks have accumulated, elongation of the rotation period may result in increases in SOC stocks by 0.24 Pg C (0.26 billion tn C) in the forest floor and by 0.37 Pg C (0.41 billion tn C) in 0–30 cm (0–12 in) depth. The increases were calculated based on SOC density and area data reported in Johnson and Kern (2003). However, it is highly unlikely that these gains can be achieved as the aim of forest management is to increase productivity of trees for timber, and not to increase SOC stocks. Until recently, forest soil management in the U.S. has been characterized primarily by inattention or risk avoidance (Williams et al. 2020b). Additional studies on the impacts of rotation cycles, harvesting and site preparation for managing SOC stocks in temperate coniferous forests are, therefore, needed.

The C impact of thinning in temperate coniferous forests in the U.S. is relatively short-term, with the greatest effects in the year of thinning (Amiro et al. 2010). Similar to boreal forests, thinning may have no overall effect on SOC stocks (Zhang et al. 2018). For example, multiple thinning did not affect SOC stocks in the forest floor and surface mineral soil (30-cm [12-in] depth) in a red pine forest in Minnesota (Jurgensen et al. 2012). The heaviest-thinned and uncut control red pine stands had higher SOC stocks in the mineral A horizon, as compared to the other studied thinning treatments. However, any SOC losses shortly after thinning may be compensated by large amounts of initial C input and forest ecosystem recovery. Thus, SOC stocks after thinning interventions do not change at later stages (Zhang et al. 2018). Further, stem-only removal for wildfire risk reduction and bio-energy production may have little impact on SOC stocks (Jurgensen et al. 2012). Otherwise, selective harvesting as part of continuous-cover forestry is considered a possible measure to reduce SOC losses (Jandl et al. 2007).

Prescribed fires burned in total 0.019 million km^2 (0.007 million mi^2) between 1992 and 2012 in Marine West Coast Forests and Northwest Forested Mountains (Balch et al. 2017). Whether this has affected SOC stocks and whether avoiding prescribed fires would have maintained temperate coniferous forest SOC stocks in the U.S. is debatable (Johnson and Curtis 2001). Fire suppression policies particularly in the western U.S.'s dry temperate forests affect biomass C storage (North and Hurteau 2011), and fire suppression activities have tended to increase forest C stocks (Domke et al. 2018), including those of the forest floor but the database is limited (Page-Dumroese et al. 2003). Fuel reduction treatments (e.g., prescribed fire and thinning) may contribute to greater long-term forest C storage (Domke et al. 2018). Fuel treatments can contribute to a decrease in fire intensity and severity, and improve forest resilience to fire, insects, and drought (Halofsky et al. 2020). However, prescribed fire appears to have little effect on SOC stocks based on the few studies specifically designed to monitor changes in SOC (Hoover 2003). Otherwise, altered, ecologically novel, high-severity fire regimes following systematic wildfire suppression may result in the reduction in SOC stocks and damp C cycling processes in the long-term (Dove et al. 2020). Overall, losses from the SOC stock can potentially be reduced by fuel management and use of prescribed fire (Nave et al. 2019a). In contrast, the potential to increase SOC stocks and enhance SOC sequestration by fire suppression and fuel reduction treatments is likely to be small (North and Hurteau 2011).

Understory vegetation in temperate coniferous forests of the U.S. can be managed to accelerate tree development and early productivity (Powers et al. 2013). However, studies on the long-term impacts of understory vegetation management on SOC stocks are scanty. For example, studies of five long-term groups in the western U.S. entirely or predominantly stocked with overstorey Ponderosa pine did not report adverse effects by vegetation management on SOC stocks under a Mediterranean climate. Otherwise, understory shrubs can profoundly affect susceptibility to wildfire and may result in losses of SOC (Nave et al. 2011; Powers et al. 2013).

Deliberate reforestation after harvest or disturbance (e.g., fire) may reduce SOC losses and enhance SOC sequestration rates compared to passive management approaches such as waiting for natural regeneration (Nave et al. 2019b). Selecting high-SOC species for reforestation may particularly reduce SOC losses (Nave et al. 2019a). Opportunities to increase SOC stocks are the introduction/favoring of tree species that are faster-growing, and/or more tolerant of heat or drought (Nave et al. 2019a). This may, however, also mean replacing coniferous with deciduous species as multi-year studies have indicated that deciduous forests continue to be a net C sink under increased temperatures and larger variability in precipitation under future climate, while the response of coniferous forests remain uncertain (Beamesderfer et al. 2020).

Whether reforestation with specific temperate coniferous species results in higher SOC stocks is unclear. Some species may be more suitable for C sequestration in stable form in the mineral soil (Vesterdal et al. 2013). Shortening the period of years to decades when the re-growing forest is a net C source to the atmosphere may enhance the SOC stock over the lifetime of the forest stand. For example, while reforestation by *Pinus taeda* L. since the 1950s in the Calhoun Experimental Forest, South Carolina, has resulted in net SOC accumulation in the topsoil, a net loss of subsoil SOC stocks was observed on a soil severely degraded by cultivation in the nineteenth and early twentieth centuries (Mobley et al. 2015). Thus, recovery times for SOC lost during cultivation may be centuries long (Nave et al. 2019b). Overall, whether SOC stocks and sequestration in temperate coniferous forests of the U.S. can be enhanced by reforestation is unclear.

While virtually no fertilizer is used in non-plantation forest systems, fertilizer applications to tree plantations are increasing (Smethurst 2010). For example, Pinus plantations in the south-eastern U.S. are commonly fertilized at establishment with N-P-K mixtures (Fox et al. 2007). About 4000 km^2 (1544 mi^2) of pine forests are fertilized each year in the Southeast (Albaugh et al. 2018). Forest fertilization, in particular, with N and its benefits for aboveground woody biomass and productivity adds substantially to SOC stocks (Johnson and Curtis 2001; Schulte-Uebing and de Vries 2018). Fertilization can also reduce SOC losses by compensating for residue removals (Nave et al. 2019a). Many perhumid rainforests on the West Coast are co-limited by N and P, and some possibly by P alone (Kranabetter et al. 2020). The potential to increase productivity and SOC stocks by fertilization with P in P-limited forests is, however, uncertain. From a C sequestration point of view, the hidden costs of increased emissions of N_2O, decreased absorption of CH_4, and GHG emissions associated with manufacture and application of the fertilizer must also be considered (Hoover 2003).

Ungulate herbivores may reduce aboveground biomass and litter C in forests but studies on their effects on SOC stocks are scanty (Mayer et al. 2020). Herbivores prefer fast-growing *vs.* slow-growing tree species (Ramirez et al. 2018), but what this means for ungulate herbivore exclusion effects on SOC stocks in temperate coniferous forests of the U.S. needs additional research.

To sum up, forest management interventions during harvesting, thinning, reforestation and prescribed burning may not have significant or lasting effects on SOC stocks in temperate coniferous forests of the U.S. (Hoover 2003). However, reducing the extent of harvested area on a landscape level to allow long-term C accumulation, N-fertilization, and introduction/favoring faster-growing trees species and those more tolerant of heat or drought are opportunities to increase SOC stocks and enhance sequestration in temperate coniferous forests of the U.S. (Nave et al. 2019a).

Temperate Broadleaf and Mixed Forest Biome

The SOC stock of managed temperate broadleaf forests is influenced by management activities, i.e., silvicultural treatments through effects of harvesting, cultivation, site preparation, fire, species conversion, reforestation/succession, and fertilization/N fixation (Hoover 2003). The management of Mediterranean forests in California is particularly driven by a focus on the natural range of variation and natural disturbance emulation concepts as well as the generation of heterogenous stand structures in montane forests (Safford and Vallejo 2019). Key SOC vulnerabilities in the temperate broadleaf forest biome in the U.S. are similar to those of temperate coniferous forests.

Natural disturbances

The SOC stock of temperate broadleaf forests in the U.S. may be altered by fire. Lightning-caused fires burned 0.009 million km^2 (0.004 million mi^2) from 1992 to 2012 in the U.S. ecoregions Northern Forests and Eastern Temperate Forests (Balch et al. 2017). Similar to temperate coniferous forests, the importance of fire for the vulnerability of SOC increases as the climate dries (Nave et al. 2019a). Low severity fire regimes are concentrated in woodlands and open forest ecosystems, dominated by oaks and certain fire-resistant pine species in California (Safford and Vallejo 2019). However, extensive drought- and beetle-driven mortality events, and century-long policy of fire suppression has resulted in fuel accumulation in some forest areas in California. This has created the potential for large, destructive forest fires. Further, temperate mixed forests (broadleaf and conifer species), which dominate the forests of California are projected to be the most vulnerable biome type to the combination of global warming and increased fire activity. However, warming induced increases in SOC loss and drought induced decreases in SOC loss may cancel each other out, resulting in no changes in SOC stocks in much of California's Mediterranean forests. Otherwise, higher fire probabilities, frequencies and severities are projected for California as are a greater annual area burned and a lengthened fire season but the effects on SOC stocks are unclear (Safford and Vallejo 2019).

Wildfires have been shown to reduce temperate forest floor C stocks to a higher degree than that for prescribed fires (−67% *vs.* −46%; Nave et al. 2011). The presence of hardwoods generally mitigates fire effects on forest floor C stocks as hardwoods produce less flammable foliage, litter, and woody detritus than softwoods. Otherwise, mineral SOC stocks in temperate forests are generally not affected by fire (Nave et al. 2011). Thus, losses of forest floor C stocks in temperate and mixed forests in the U.S. may be reduced by increasing the frequency of prescribed fires and the abundance of hardwood species. For example, the increased use of prescribed fire in Mediterranean forests in California has recently been promoted to support forest resilience (Safford and Vallejo 2019).

Insect infestations affect temperate broadleaf forests in the U.S. but the C sink may only be reduced in the year of infestation, and generally recovers in subsequent years (Amiro et al. 2010). For example, hemlock trees may be infested and killed by the hemlock woolly adelgid (*Adelges tsugae* [Annand, 1928]) in the eastern U.S. (Williams et al. 2016). This process transfers live biomass to coarse woody debris but the loss of C by decomposition may be compensated by released hardwood growth and enhanced SOC stabilization in association with soil minerals (Fraterrigo et al. 2018; Raymer et al. 2013). Thus, temperate broadleaf forest ecosystems infested with the hemlock woolly adelgid may not loose C.

Invasion of emerald ash borer (EAB: *Agrilus planipennis* Fairmaire, 1888) in the Great Lakes region of the U.S. has resulted in high mortality of Fraxinus species (Williams et al. 2016). However, the associated reduction in NPP may have been partially offset in the short term by compensatory growth of Acer and Ulmus species. Overall, the loss of ash trees may not be a primary driver of forest change over the next century at landscape scales (Olson et al. 2021). Otherwise, Matthes et al. (2018) suggested that SOC losses may occur depending on soil properties, and, thus, forest SOC stocks may decline in the long-term following EAB infestations (Williams et al. 2016). Forest tent caterpillars (*Malacosoma disstria* Hübner, 1820), spruce budworms (*Choristoneura* spp.) and gypsy moths (*Lymantria dispar dispar* [Linnaeus, 1758]) may cause defoliation in northeastern U.S. forests (Williams et al. 2016). However, impacts of these insects on forest floor C and SOC stocks vary (Hicke et al. 2012). Thus, loss of SOC from some EAB infected temperate broadleaf forests may be reduced by insecticide treatment and biological management. Forest pests and diseases particularly complicate management of Mediterranean forests in California (Safford and Vallejo 2019).

Similar to both boreal and temperate coniferous forests, the effects of windthrow on forest floor and mineral SOC stocks in temperate broadleaf forests may be small if disturbance occurs within the range of natural variability (Liechty et al. 1997). However, in extreme cases blowdowns of whole stands occur from microbursts, tornadoes, cyclones or hurricanes. Tropical cyclones are a principal agent of catastrophic wind damage in temperate forests of the eastern U.S., with dramatic impacts on forest structure and functioning (Fisk et al. 2013). Even for the more extreme wind events, the net effect on forest C sequestration is close to neutral at large scales because disturbance emissions are largely offset by the regrowth legacy stimulated by past events (Williams et al. 2016). Options to reduce wind-induced SOC losses may be retaining residues, fertilization or reforestation of wind-damaged forests (Nave et al. 2019a).

Ice storms are common to some temperate forests in the U.S. (Changnon 2003). Large amounts of woody debris on the forest floor may be created by ice storms. However, this may not translate into increased SOC stocks as decomposition may also be enhanced by more open canopies (Adams et al. 2019). Retaining residues, fertilization or reforestation of ice storm-damaged forests may be opportunities to reduce SOC losses.

Stand management

Forest floor C stocks are generally more vulnerable to harvest-induced losses than mineral SOC stocks (Nave et al. 2010). In temperate broadleaf forests growing on Podzols, both forest floor and organo-mineral subsoil C stocks may be destabilized by harvesting (Nave et al. 2019a). However, only intensive harvesting may generally result in losses from the SOC stock of temperate broadleaf forests with amounts comparable to those from temperate coniferous forests (Achat et al. 2015). For example, clear cutting resulted in SOC losses from strongly-mineral bound and stable C pools in a northern hardwood forest in the U.S. compared to an uncut mature forest (Lacroix et al. 2016). In contrast, Petrenko and Friedland (2015) found no differences between SOC stocks in >100-years-old and harvested hardwood forests in the Northeastern U.S. However, a long-term decrease in SOC stocks may occur when primary forests are logged (Dean et al. 2017). Thus, harvest intensity can be adjusted to reduce losses from SOC stocks in temperate broadleaf forests. Similar to coniferous forests, rotation period lengths can also potentially be adjusted to accumulate SOC and enhance SOC sequestration in temperate broadleaf forests in the U.S. but data are scanty.

Thinning interventions may not change SOC stocks in temperate broadleaf forests similar to that of boreal and temperate coniferous forests (Zhang et al. 2018). For example, multiple thinning interventions did not affect SOC stocks in the forest floor and surface mineral soil (30-cm [12-in] depth) in a northern hardwood forest in Wisconsin (Jurgensen et al. 2012). However, SOC dynamics may be affected by thinning. For example, soil respiration was increased by 35.6% and 9.3% in temperate broadleaf and temperate mixed forests, respectively (Zhang et al. 2018). Expansion of thinning interventions in Mediterranean forests has recently been promoted in California (Safford and Vallejo 2019), but effects on SOC stocks are uncertain.

Between 1992 and 2012, prescribed fires burned 0.04 million km^2 (0.02 million mi^2) in Northern Forests and Eastern Temperate Forests (Balch et al. 2017). How this has affected long-term SOC sequestration and stocks, and the possibility to manage SOC stocks by prescribed fires in U.S. temperate broadleaf forests are, however, unclear.

The forest floor C and SOC stocks under different broadleaf tree species in temperate regions are variable (Vesterdal et al. 2013). For example, Fagus accumulates more C in forest floors than Fraxinus, Acer, and Tilia, whereas Quercus spp. has intermediate forest floor C stocks. More than 2-fold differences between the species with least and most forest floor C stocks are observed. However, mineral SOC stocks may be less affected by tree species although there is a tendency of more SOC under certain hardwood tree species, i.e., Fraxinus, Acer, Tilia and Ulmus than under Fagus (Vesterdal et al. 2013).

Temperate broadleaf trees in the U.S. associated with ECM had higher SOC stocks in the topsoil while overall SOC stocks to 1-m depth were higher under trees associated with AM (Craig et al. 2018). Further, root-derived C was 54% greater in AM versus ECM-dominated temperate forest plots, and twice as much root-derived C was found in mineral-associated OC in AM compared to ECM plots (Keller et al.

2021). This implies that the SOC stocks in AM-dominated soils were more stabilized compared to that in ECM-dominated soils. Thus, reforestation with trees associated with AM fungi may increase SOC stocks and enhance SOC sequestration probably due to higher root-derived soil C inputs. However, effects of tree species may be masked by soil conditions and site-related factors (Craig et al. 2018). Nevertheless, losses of SOC can potentially be reduced and SOC stocks increased in temperate coniferous and mixed forests in the U.S. by tree species selection during reforestation, including planting N-fixing trees (Hoover 2003). High-SOC species and those that are faster-growing and more tolerant of heat or drought should be considered for reforestation (Nave et al. 2019a). Increases in SOC stocks in 0–30 cm (0–12 in) depth by 0.96 Mg C ha^{-1} y^{-1} (0.43 tn C ac^{-1} y^{-1}) have been reported in warm temperate climates where plantations are mainly composed of deciduous species (Wang and Huang 2020). However, management decisions are guided by the need to increase timber production and not to enhance SOC sequestration.

Forest fertilization may result in increases in SOC stocks of temperate broadleaf forests in the U.S., but the hidden costs of fertilization for C sequestration need also to be considered (Johnson and Curtis 2001; Hoover 2003). N-fertilization, in particular, is an opportunity to increase aboveground woody biomass and SOC stocks (Nave et al. 2019a; Schulte-Uebing and de Vries 2018). Some temperate broadleaf forests in the U.S. are showing changes in line with moving from an N-limited to a P-limited system caused by increases in atmospheric N deposition along with Mg, K, and Ca deficiencies (Crowley et al. 2012). Whether fertilization with P, Mg, K and Ca increases tree growth and subsequently SOC stocks in temperate broadleaf forests is not known.

Forest liming studies have been conducted in conifer stands and mixed hardwood forests in the U.S., but have been most common in stands of sugar maple (*Acer saccharum* Marshall; Melvin et al. 2013). However, very few studies have investigated how SOC sequestration and stocks are influenced by forest liming. For example, forest floor C stocks were higher in limed plots in a mixed hardwood forest in New York compared to that of the control (68.1 Mg C ha^{-1} *vs.* 30.7 Mg C ha^{-1} [30.4 tn C ac^{-1} *vs.* 13.7 tn C ac^{-1}]; Melvin et al. 2013). Primarily liming-induced reduction in decomposition rates may account for much of the observed increase in forest floor C stocks. However, additional research is needed to identify the primary drivers of liming effects as mixed hardwood forest of northeastern U.S. have historically experienced also high acid deposition rates, declines in soil pH and exchangeable Ca availability (Melvin et al. 2013). High-dose liming of acidic N-rich forest soils can result in long-term decreases in SOC stocks (Persson et al. 2021).

Excluding ungulate herbivores in temperate broadleaf forests may affect SOC stocks but the database is limited (Mayer et al. 2020). For example, exclusion of elk [*Cervus canadensis* (Erxleben, 1777)] significantly increased SOC in regenerating willow (*Salix* spp.) and aspen stands in the Rocky Mountains and in Arizona, respectively (Gass and Binkley 2011; Stritar et al. 2010). More studies on ungulate herbivory exclusion effects on SOC in broadleaf temperate forests of the U.S. are needed.

To sum up, similar to temperate coniferous forests in the U.S., key SOC vulnerabilities in temperate broadleaf forest in the U.S. are associated with harvest practices and fire. Likewise, opportunities to increase SOC stocks and enhance sequestration include reducing the extent of harvested area on a landscape level to allow long-term C accumulation, N-fertilization, and introduction/favoring faster-growing trees species and those more tolerant of heat or drought (Nave et al. 2019a).

Tropical Forest Biome

The SOC stock of tropical moist forests in the U.S. is vulnerable to forest degradation (e.g., due to overgrazing, fuelwood extraction or high-grade selective harvesting), deforestation (including forest conversion to croplands or pastures, mining activities), and native forest conversion to tree plantations (Nave et al. 2019a). Management practices may particularly affect the SOC stocks of short-rotation plantations on tropical moist forests. Further, many tropical moist forests in the U.S. are in some stage of recovery from past human disturbance with much regenerating into secondary forests and a smaller area supporting tree plantations (Silver et al. 2003). The SOC stocks of tropical moist forests may also be affected by natural disturbances similar to that of tropical dry forests (TDFs).

Natural disturbances

Tropical moist forests in coastal areas of the U.S. and the U.S.-affiliated Atlantic and Pacific Islands may be affected by cyclonic storms (hurricanes, typhoons and cyclones; Boose et al. 1994; Oswalt et al. 2019). Tropical cyclones shape the species composition and structure of tropical forests, in particular, that of tropical rainforests (Corlett and Primack 2011). Hurricanes may cause defoliation, and substantial losses of branches and apical dominance causing an extensive reduction in canopy cover. Higher mortality is associated with rugged terrain and higher elevations due to uprooting and landslides (Oswalt et al. 2019). Both maximum wind speeds and hurricane-related rainfall may cause forest damage (Hall et al. 2020).

Hurricanes in tropical moist forests in Puerto Rico may cause increased soil and groundwater ammonium and nitrate concentrations, and increased groundwater dissolved organic N concentrations (Jaramillo and Murray-Tortarolo 2019). Eventually, values return to pre-hurricane levels. Otherwise, nutrient concentrations and fluxes in storm runoff or in stream flows have shown higher nitrate and dissolved and particulate P concentrations after hurricane disturbance in tropical moist forests of Puerto Rico. This intense disturbance of nutrient dynamics may affect ecosystem productivity (Jaramillo and Murray-Tortarolo 2019), and eventually SOC stocks.

Individual hurricane events may transfer a large amount of C to the soil. For example, hurricane Katrina transferred 105 Tg C (116 million tn C) from live to dead forest C pools (Chambers et al. 2007). In 2017, hurricanes María and Irma resulted in the loss of 10 Tg C (11 million tn C) of forest aboveground biomass across Puerto Rican forests (Hall et al. 2020). Average annual biomass kill from hurricanes are estimated to be of similar magnitude to total emissions from wildfire but with very large interannual variability ranging from 0 to as much as 200 Tg C (221 million tn C) in any given year (Williams et al. 2016). However, the net effects on C sequestration and possibly SOC stocks are close to neutral at large scales because disturbance emissions are largely offset by the regrowth legacy stimulated by past events. Active reforestation, facilitation of N-fixer establishment, fertilization, selection of high-SOC species/genetic families for plantations, and retaining residues during salvage logging may be opportunities to increase SOC stocks of hurricane-damaged tropical U.S. forests (Nave et al. 2019a).

Changes in drought frequency and magnitude, extreme rainfall events, and atmospheric N deposition may alter SOC dynamics in tropical moist forests of the U.S. (Cusack and Marín-Spiotta 2019). The SOC stocks are particularly vulnerable to changes in rainfall. Thus, active reforestation with more drought resistant or drought tolerant species may be options to reduce SOC losses or increase SOC stocks in tropical moist forests affected by drought events (Cusack and Marín-Spiotta 2019).

Lightning-caused fires burned 0.004 million km^2 (0.001 million mi^2) from 1992 to 2012 in the CONUS ecoregion Tropical Wet Forests (Balch et al. 2017), but the effects on SOC stocks are unclear. The SOC stocks of TDFs in the U.S. are susceptible to fire aside to forest degradation, deforestation and grass invasion (Nave et al. 2019a). However, the TDF biome is remarkably resistant to fire occurrences (Jaramillo and Murray-Tortarolo 2019).

Stand management

Reforestation of mature forest land following abandonment of agricultural activity can result in increases in SOC stocks of tropical moist forests but the database is limited (Silver et al. 2003). In Puerto Rico, for example, secondary forests with new tree species combinations and forest plantation had higher SOC stocks in 0–10 cm (0–3.9 in) depth than mature forests whereas SOC stocks where highest in mature forests in 0–25 cm (0–9.8 in) depth. The SOC can accumulate rapidly with reforestation in surface soils but it may take decades until a new steady-state in profile SOC stocks is reached (Silver et al. 2003). Otherwise, SOC losses during reforestation can be reduced by preventing grass invasions, establishing agroforestry systems, and rotation lengths sufficiently long to recover SOC stocks (Nave et al. 2019a). The selection of high-SOC species in plantations and mixed-species plantations are other options to reduce SOC losses. The SOC of Ferralsols in tropical moist forests can turn over very rapidly, and, thus, sustained C inputs are needed for this soil group to maintain SOC stocks (Nave et al. 2019a).

Reducing physical disturbance during stand management can maintain aggregate-protected SOC in TDFs growing on Lixisols as these soils have limited structural stability (Nave et al. 2019a). Agroforestry can mitigate SOC losses from TDFs growing on Lixisols. The SOC stocks of subtropical humid forests on Alisols are also susceptible to erosion, and physical disturbance should be reduced during stand management. Mulching and organic amendments can maintain SOC stocks in steeply sloping Kastanozems of subtropical mountain forests (Nave et al. 2019a). As human exploitation of TDFs is higher than that of other tropical forest biomes (Jaramillo and Murray-Tortarolo 2019), extraction of fuelwood and thin stems from TDFs should be reduced to maintain SOC stocks.

Between 1992 and 2012, prescribed fires burned 0.004 million km^2 (0.001 million mi^2) in Tropical Wet Forests in CONUS (Balch et al. 2017). Whether fires in tropical forests can be managed to enhance SOC stocks and sequestration is, however, unclear. Specifically, effects on SOC in fire-sensitive systems of shorter return intervals and more intense fires requires greater research efforts (Mayer et al. 2020).

Judicious management of land uses that can lead to forest degradation, such as grazing and fuelwood harvesting are among options to reduce SOC losses in TDFs (Nave et al. 2019a). In disturbed TDFs, preventing further deterioration of site productivity through fire management, and the exclusion of grasses that increase fire susceptibility aids in maintaining forests, along with supplemental planting to maintain canopy cover. Reforestation including that with drought resistant or drought tolerant plant species can be a successful way to increase SOC stocks in TDFs as can be the facilitation of N-fixer establishment. Further, addition of surface organic amendments and residue retention can retain soil water, which can enhance survivorship of planted trees during drought events, and in turn enhance stand productivity and SOC stocks in TDFs (Nave et al. 2019a).

Opportunities to increase SOC stocks in developing secondary tropical moist forests include active reforestation, facilitation of N-fixer establishment, reforestation of degraded soils with high-SOC species, deliberate reforestation of eroded slopes, and erosion control (Nave et al. 2019a). For example, including N-fixing trees in reforestation projects can enhance biomass C stocks in tropical moist forests (Levy-Varon et al. 2019). This increase in C capture may eventually result in increases in SOC stocks (Mayer et al. 2020). However, soil P and Mo, and frequency of disturbance can limit N-fixation and tree growth in tropical moist forests. Otherwise, P additions to tropical moist forests may not affect SOC stocks as indicated by their small effects on decomposition and microbial respiration (Camenzind et al. 2018). Similarly, aboveground woody biomass may not respond to P addition in tropical forests (Schulte-Uebing and de Vries 2018). The potential impact of N fixation on C uptake by mature tropical moist forests is also uncertain (Levy-Varon et al. 2019).

Fertilization either with N or base cations may increase SOC stocks in tropical moist forests while plantation SOC stocks may be increased through selection of high-SOC species or genetic families (Nave et al. 2019a). However, soils of tropical moist forests may have overall a limited capacity to sequester additional C inputs (Sayer et al. 2019). Further, the aboveground woody biomass of tropical forests does not respond to N additions (Schulte-Uebing and de Vries 2018). Nevertheless, SOC sequestration can be stimulated in mature tropical forests under excess N deposition but this sink may not be sufficient to compensate for all the GHG emissions derived from N-fertilizer production, transport and application (Lu et al. 2021).

Exclusion of large mammal and ungulate herbivory in tropical moist forests may have only small effects on SOC stocks (Tanentzap and Coomes 2012). Otherwise, to reduce SOC losses from domestic ungulate herbivory, Schulz et al. (2016) recommended grazing management that considers compulsory rotation systems.

To sum up, SOC losses from tropical moist forests in the U.S. can potentially be reduced by fire management, prevention of grass invasions, selection of high-SOC species for plantations, mixed-species plantations, reforestation of burned areas, and grazer density control. Increases in SOC stocks may be achieved by reforestation, facilitation of N-fixer establishment, control of soil erosion, selection of high-SOC species or genetic families on degraded soils and for plantations, and retaining logging residues (Nave et al. 2019a). Management of fire and selecting drought-adapted species may be promising for managing SOC stocks and sequestration in TDFs.

3.5.2 Grassland and Shrubland Biomes

Temperate Grasslands, Savannas, and Shrublands Biome

Mineral and organic soil of grassland remaining grassland lost SOC in the year 2019 (Table 3.1; U.S. EPA 2021). Weather patterns and land management strongly affect productivity and C cycling in temperate grasslands (Pendall et al. 2018; U.S. EPA 2021). Globally, the GHG sink strength of grassland and grazing land is reduced following intensive management (i.e., excreta N deposition, manure/fertilizer N application and manure N management practices) and increased livestock density (Dangal et al. 2020). Higher N inputs in the form of excreta, manure and fertilizer are the dominant drivers of increases in grassland net C uptake. Regional and grassland-specific differences in GHG fluxes should guide future management decisions (Dangal et al. 2020). In 1750–2012, direct human management activities may have caused grasslands to switch from a sink to a source of GHGs, because of increased livestock numbers and accelerated conversion of natural lands to pastures (Chang et al. 2021). However, climate change drivers supported a net SOC sink, mainly by increased productivity of grasslands due to increased atmospheric CO_2 and N deposition. Globally, the net radiative forcing of grasslands is currently close to neutral. In the future, optimized grazing, pasture improvement and restoration of degraded pasture will be critical to enhance grassland SOC sequestration (Chang et al. 2021).

Temperate grasslands in the U.S. are estimated to have been small net C sinks in recent years based on flux tower observations and biogeochemical modelling but also net C sources in some years as inter-annual variability in grassland C fluxes is large (Chang et al. 2021; Liang et al. 2020). Decreased livestock numbers in grasslands in the central U.S. have resulted in enhanced SOC storage (Chang et al. 2021). However, in 2019 mineral and organic temperate grassland soils in CONUS and Hawaii excluding those of Alaska (covering 0.5 million km^2 [0.2 million mi^2]) emitted 3.95 Tg C (4.35 million tn C; Land-use category 'Grassland Remaining Grassland'; U.S. EPA 2021). There has been considerable variation in SOC stocks between 1990 and 2019, with average annual change of 2.4 Tg C (2.6 million tn C). Specifically, there is a large discrepancy between modeling estimates and empirical, data-driven values that contributes to uncertainty in the assessment of changes in grassland SOC stocks. For example, average annual net temperate grassland C uptake in the U.S. in 2000–2006 has been estimated at 13.2, 266.2 and 104.8 Tg C

y^{-1} (14.6, 293.4 and 115.5 million tn C y^{-1}) based on inventory analysis, atmospheric inversion models and land-surface models, respectively (Pendall et al. 2018).

The temperate grassland SOC stock is particularly sensitive to management, and may increase by improved grazing management, fertilization, irrigation, increasing species diversity, and sowing legumes and improved grass species (Conant et al. 2017). However, any increase in SOC stock following N additions (animal excretions, fertilizers) must be evaluated in relation to increases in N_2O emissions (Du et al. 2021), aside GHG emissions associated with fertilizer production, transport and application. Godde et al. (2020) cautioned against unrealistic expectations of SOC management in grazing systems as a mitigation strategy. For example, higher plant diversity in temperate grasslands may increase rhizosphere C inputs into the soil microbial community resulting in increased SOC storage (Lange et al. 2015). Otherwise, mixture effects on grassland SOC stocks were not found by Chen et al. (2020a), probably because experimental durations were too short. Grazing in combination with fertilization may result in higher SOC stocks in 0–10 cm (0–4 in) depth than without nutrient additions (Sitters et al. 2020). There is, however, a limited number of observations and practices occurring in particular contexts or regions, and SOC sequestration rates in grazing lands are highly context-dependent (Godde et al. 2020).

The temperate grassland SOC stock can be rapidly lost through processes such as soil disturbance, vegetation degradation, fire (wildfire or managed fire), erosion, nutrient shortage and water deficit, and the inappropriate management of livestock grazing (McSherry and Ritchie 2013; Soussana and Lemaire 2014). Otherwise, herbivore exclusion may not result in SOC losses under ambient nutrient conditions (Sitters et al. 2020). Disturbance of the continuous temperate grassland vegetation cover may particularly contribute to increased losses of SOC (Soussana et al. 2004). However, it is not fully understood how the SOC stocks of temperate grassland respond to land and soil degradation (Dlamini et al. 2016; Sitters et al. 2020).

Grassland SOC stocks may also be lost by conversion to other land uses. For example, conversion of grassland to corn and soybeans in 12 U.S. Midwestern states (Illinois, Indiana, Iowa, Kansas, Michigan, Minnesota, Missouri, Nebraska, North Dakota, Ohio, South Dakota, Wisconsin) resulted in an estimated SOC loss by 5.6% between 2008 and 2016 relative to that of existing cropland (Zhang et al. 2021). Over 20,000 km^2 (7722 mi^2) of grassland were converted to crop production in these states.

Natural disturbances

Fires are a common natural phenomenon in U.S. grasslands (Anderson 2006). From 1992 to 2012, lightning-started fires accounted for 11, 39, 43, 79 and 60% of the area burned in CONUS ecoregions with significant grassland cover, i.e., Mediterranean California, Great Plains, Southern Semiarid Highlands, North American Desert and Temperate Sierras, respectively (Balch et al. 2017). This corresponded to a total area of 0.13 million km^2 (0.05 million mi^2) burned following lightning. Fire can substantially alter soil characteristics and processes during burning, and the post-fire recovery period (Santín and Doerr 2016). For example, soil

CO_2 fluxes increased by 20–55% following fire in a tallgrass prairie (Knapp et al. 1998). Otherwise, the temperate grassland SOC contents can increase where post-fire erosion is limited, and where extremely severe or frequent fires are absent. However, the overall effect of long-term pyrogenic OM inputs to the soil from fire on temperate grassland SOC stocks depends on how much C is lost in combustion and by erosion (Soong and Cotrufo 2015).

Management

The SOC stocks of grazing lands in the U.S. can be increased by management intensification including species placement, fertilization (i.e., with N, P, K, micronutrients; Ochoa-Hueso et al. 2020), species richness, addition of legumes and defoliation management with benefits for SOC depending on the baseline management intensity (i.e., prior to intensification) and the grassland age (Fay et al. 2015; Sarkar et al. 2020; Sollenberger et al. 2019). For example, SOC stocks in 0–30 cm (0–12 in) depth under sown bahiagrass (*Paspalum notatum*) pasture managed more intensively (fertilizer, rotational stocking, and greater stocking rate) in Florida were 21 Mg C ha^{-1} (9 tn C ac^{-1}) higher than those for the original native rangeland (Adewopo et al. 2014). Forage and biomass plantings may increase SOC stocks by 0.02–0.17 Mg C ha^{-1} y^{-1} (0.01–0.07 tn C ac^{-1} y^{-1}) on average over 20 years based on estimates by the COMET-Planner (Swan et al. 2015). Range planting, i.e., the establishment of adapted perennial or self-sustaining vegetation such as grasses, forbs, legumes, shrubs and trees may result in SOC plus woody biomass C stock increases by 0.22–0.35 Mg C ha^{-1} y^{-1} (0.09–0.14 tn C ac^{-1} y^{-1}; Swan et al. 2015).

Grazing systems may have the potential to absorb the largest amount of additional tree cover (Chapman et al. 2020). However, increases in SOC stocks with increases in plant species richness including the presence of legumes and trees at temperate grassland sites in the U.S. are not always observed. There have also been few studies on the effects of legume introduction on SOC accumulation, and it is assumed that improvements depend in part on the saturation of the mineral-associated SOM pool (Sollenberger et al. 2019). AM interactions may be critical for enhancing SOC sequestration following the introduction of legumes in forage rotations (Sarkar et al. 2020).

Grassland species composition is heavily impacted by the introduction and spread of nonnative plant species, especially in the tallgrass prairie of the central U.S. (Wilsey et al. 2020). The nonnative dominated grasslands and savannas may differ in total soil C stocks from native-dominated areas. For example, no changes in SOC stocks to 1-m depth (39-in) 8 years after introduction of nonnative grass species have been observed but decreases in carbonate-C, and, thus, total soil C stocks compared to those under native grass species (Wilsey et al. 2020).

DeLuca and Zabinski (2011) estimated that the combined SOC losses from tall-, mixed-, and shortgrass prairies in the Great Plains of North America by conversion to cropland use may amount to 5 Pg C (5.5 billion tn C). Implementation of alternative agricultural production systems that more closely emulate the biodiversity, phenology, and biogeochemical processes associated with native prairie ecosystems such as perennial polycultures may restore some of the SOC stock lost from the Great Plain prairies (Crews et al. 2018).

Ogle et al. (2004) estimated that management of U.S. temperate grasslands including irrigation and seeding legumes may increase SOC storage by sequestering 0.5 and 0.4 Mg C ha^{-1} y^{-1} (0.20 and 0.16 tn C ac^{-1} y^{-1}) in pastures and rangelands, respectively. Implementing an additional improvement (high input) on all managed temperate grasslands would sequester 0.7 and 0.8 Mg C ha^{-1} y^{-1} (0.28 and 0.32 tn C ac^{-1} y^{-1}) on pastures and rangelands, respectively. Otherwise, converting managed temperate grasslands in the U.S. from degraded to nominal conditions would have only a small effect on SOC storage, sequestering C at a rate of about 0.1 Mg C ha^{-1} y^{-1} (0.04 tn C ac^{-1} y^{-1}; Ogle et al. 2004). In total, management improvements would increase sequestration in pastures and rangelands to a rate of 40–46 Tg C y^{-1} (44–51 million tn C y^{-1}) which is similar to the rate for U.S. grazing lands of 54 Tg C y^{-1} (60 million tn C y^{-1}; Follett et al. 2001). However, these potential rates are relevant for about 20 years only as grassland management practices have a limited time frame over which C sequestration would occur (Conant et al. 2001). Further, the relationship between grassland age and SOC accumulation is challenging to predict (Sollenberger et al. 2019).

In 2015, about 0.06 million km^2 (23,000 mi^2) of non-cultivated cropland including permanent hayland were irrigated in the U.S. (USDA 2018). Irrigation of moisture-limited grasslands such as those located in arid and semi-arid regions can increase biomass production and, thus, soil C input (Hristov et al. 2018). For example, yields of hay [combination of alfalfa hay (*Medicago sativa* L.) and clover-timothy (*Trifolium* L.-*Phleum pratense* L.) mixtures, bermuda grass (*Cynodon dactylon* (L.) Pers.), prairie hay, etc.] in CONUS were increased by 137% at irrigated fields compared to those without irrigation in 1950–2016 (Lu et al. 2020). However, the improved soil moisture with irrigation may not only lead to increased plant-derived soil C input but also to increased decomposition of SOC by promoting decomposer activity (Trost et al. 2013). Thus, the net effects of irrigation on SOC depend on climate, initial SOC stocks, fertilization, soil structure and decomposer community. For example, SOC stocks to 1-m (39-in) depth were 59 Mg C ha^{-1} (26 tn C ac^{-1}) under native sagebrush (*Artemisia tridentata* Nutt.) vegetation but 101 Mg C ha^{-1} (45 tn C ac^{-1}) under irrigated pasture in a semiarid-arid region in Idaho after 30 years of irrigation (Entry et al. 2004). However, long-term experiments are scanty, and the database about irrigation effects on U.S. grassland SOC stocks is limited (Conant et al. 2017). Similar to irrigation of cropland, optimization of grassland irrigation is needed to reduce CH_4 emissions and net global warming potential (GWP; Sapkota et al. 2020).

In the short term within years, management changes have little effect on C storage in semiarid rangelands like the western U.S. Great Plains (Sanderson et al. 2020). Here, weather can dominate changes in SOC stocks. The potential to increase SOC stocks due to management of semiarid rangelands is also small on a per unit area basis. In the western Great Plain rangelands, SOC stocks may increase by 4.5% by grazing for every 100 mm (4 in) decrease in precipitation compared to un-grazed pasture controls. At 538 mm (21 in) of precipitation, the effect of grazing on SOC stocks may switch from positive to negative, i.e., grazing-induced SOC losses may occur with further increases in precipitation. However, effects of grazing management on SOC stocks are generally small in this region, and keeping semiarid rangelands intact is the most important management goal for maximizing SOC stocks. For long-term SOC increases, livestock numbers should be adjusted in response to fluctuating forage production conditions (Sanderson et al. 2020).

Human-caused fires are an important grassland management practice to remove standing dead vegetation and improve forage for grazing livestock (U.S. EPA 2021). From 1992 to 2012, human ignitions accounted for 89, 61, 57, 21 and 40% of the area burned in the ecoregions Mediterranean California, Great Plains, Southern Semiarid Highlands, North American Desert and Temperate Sierras, respectively (Balch et al. 2017). In total, 0.10 million km^2 (0.04 million mi^2) burned in the ecoregions during 1992–2012 following human ignitions. Humans are also introducing fire-promoting invasive grasses across U.S. ecoregions which increases fire occurrence and frequency at regional scales (Fusco et al. 2019). Among the regions affected by this development are pine savannahs in the southeast, deserts in the southwest, and semiarid shrublands in the Great Basin.

In recent decades, prescribed fires became increasingly important to meet specific grassland management objectives (Santín and Doerr 2016). Prescribed uses of fire include under-burning or surface fires in grasslands where fire burns ground and understory fuels but does not affect the overstory. However, fire effects on soils including SOC stocks may be negligible or very limited unless the accumulation of ground fuel available for burning is high (Santín and Doerr 2016). The changes in deep SOC stocks to multi-decadal changes in fire frequencies are generally not studied. In contrast, Pellegrini et al. (2021) reported that the assumption that SOC stocks below 20-cm (8-in) depth are not responsive to changes in fire frequency in temperate savannas is questionable. Specifically, the difference in SOC losses between intermediately burned plots versus most frequently burned plots in a temperate oak savanna in Minnesota more than doubled when the 100-cm (39 in) soil depth was considered rather than the 0–20 cm (0–8 in) soil depth alone (losses of 98.5 Mg C ha^{-1} [43.9 tn C ac^{-1}; −76%] and 42.3 Mg C ha^{-1} [18.9 tn C ac^{-1}; −68%] in the 100-cm [39-in] and 0–20 cm [0–8 in] depth, respectively). Lower inputs of root C into the deep SOC stock may be the primary contributor to lower SOC stocks with increased burning (Pellegrini et al. 2021). Overall, this sensitivity of the deep SOC stock to changes in fire frequency may affect the potential to increase temperate grassland SOC stocks in the U.S. by fire management.

Practices that decrease biomass removal from U.S. grazing lands by reducing the number of grazing animals such as prescribed grazing have the potential to increase SOC stocks on average by 0.17–0.44 Mg C ha^{-1} y^{-1} (0.07–0.18 tn C ac^{-1} y^{-1}) over 20 years (Swan et al. 2015). However, entirely excluding grazing animals is not beneficial for SOC stocks as grazed pastures have higher SOC stocks than ungrazed pastures (Sollenberger et al. 2019). Thus, temperate grassland management should include low rather than high stocking rates to support higher SOC stocks. Site-specificity in SOC response to grazing intensity may also exist (Sollenberger et al. 2019). For example, adaptive multi paddock (AMP) grazing that utilizes short-duration rotational grazing at high stocking densities increased SOC stocks by 13% (i.e., 9 Mg C ha^{-1} [4 tn C ac^{-1}]) to 1-m (39-in) depth in the southeastern US (Mosier et al. 2021). However, effects of AMP on SOC stocks vary by site (Gomez-Casanovas et al. 2021). Globally, heavy grazing reduced SOC at 0–10 and 10–30 cm (0–4 in and 4–12 in) depths (Lai and Kumar 2020). A reduction of SOC at 10–30 cm was also observed under moderate grazing. In contrast, light grazing increased SOC at 0–10 cm depth. Further, cattle grazing impacts on SOC were higher than sheep (*Ovis aries* Linnaeus, 1758) grazing (Lai and Kumar 2020).

Grazing management can result in increases in SOC stocks (Pendall et al. 2018). Compared to mowing, for example, higher and more diverse C input under grazing may lead to enhanced substrate availability and, thus, more efficient microbial functioning, which may favor SOC sequestration through formation of microbial products (Gilmullina et al. 2020). Low to moderate livestock stocking rates, in particular, are associated with higher SOC stocks compared to those under high stocking rates (Sollenberger et al. 2019). Globally, grazing significantly also increased plant richness compared to grazing exclusion, especially in wet grasslands, while effects of the duration of grazing exclusion were stronger in dry areas (Gao and Carmel 2020). These changes in plant richness were correlated with changes in SOC showing a hump-shaped pattern.

The goal of traditional grazing management is the production of livestock commodities. To adapt to changing climatic conditions, adaptive grazing management (i.e., adjustment in stocking rates and season of use) instead of a fixed management system might be better to improve grassland productivity (Zhou et al. 2019a). In contrast, the goal of targeted grazing is to apply defoliation or trampling to achieve specific vegetation management objectives (Bailey et al. 2019). Targeted grazing includes practices to optimize the timing, frequency, intensity, and selectivity of grazing (or browsing) in combinations that purposely exert grazing/browsing pressure on specific plant species or portions of the land. While the aim of traditional, adapted and targeted livestock grazing management is to sustain soil health, ultra-high stocking density is perceived as a strategy to increase SOC stocks but evidence of its effects is scanty (Bailey et al. 2019). Similarly, rotational grazing is thought to increase SOC stocks but there may be no plant production benefits, and pasture- and rangeland SOC responses in the U.S. to grazing management are generally variable (Briske et al. 2008; Derner and Jin 2012; Franzluebbers et al. 2012).

While poor grazing management has been associated with decreases in SOC stocks, improvements in grazing management may result in increases but this has been not always observed (Pendall et al. 2018). For example, rotational grazing did not result in discernable differences in plant biomass productivity, soil bulk density and SOC fractions up to 30-cm (12 in) depth compared to continuous grazing treatments in a tallgrass prairie in central Oklahoma (Franzluebbers et al. 2019; Steiner et al. 2019). Globally, continuous grazing is detrimental to SOC stocks compared to rotational or no-grazing (Byrnes et al. 2018). Otherwise, Ritchie (2020) reported that short-duration, high stocking density grazing schemes can increase SOC stocks at higher mean annual temperature and precipitation, in particular, in temperate systems. There have been few studies on the impact of cattle stocking rate on long-term changes in pasture SOC stocks in the eastern U.S. (Franzluebbers et al. 2012).

Fertilization can increase plant biomass and forage production and subsequently SOC stocks in managed pastureland if soil fertility is limited (Franzluebbers et al. 2012). Specifically, N additions to grassland soils that previously received little or no additional N may contribute to SOC sequestration (Huang et al. 2020). The N-induced productivity gain may further be promoted by liming of low diversity grasslands to reduce the negative impacts of soil acidification on productivity while maintaining high diversity in more species-rich grasslands may help ensure productivity returns (Kimmel et al. 2020). Improved N management by partial substitution of animal manure or compost for synthetic N fertilizer may also increase SOC stocks in managed pastures (Swan et al. 2015). For example, increases in SOC stocks by 2.6 Mg C ha^{-1} (1.1 tn C ac^{-1}) with high compared to low mineral fertilization rates, and by 5.7 Mg C ha^{-1} (2.3 tn C ac^{-1}) with broiler litter addition than without have been reported (Franzluebbers et al. 2012). However, C addition with broiler litter does not necessarily result in increased pasture SOC stocks. Otherwise, N fertilization had no effects on SOC sequestration in perennial-grass-based pastures in Pennsylvania (Skinner 2013). Thus, the potential to increase pasture SOC stocks by mineral and organic fertilizer management in the U.S. is uncertain.

Exogenous OM additions can stimulate temperate grassland productivity, and increase SOC stocks beyond the actual tonnage of OM applied (Bossio et al. 2020; Sarkar et al. 2020). For example, compost amendments to grasslands in California were estimated to increase active SOC stocks in 0–20 cm (0–8 in) depth by 0.14, 0.11 and 0.04 Mg C ha^{-1} y^{-1} (0.06, 0.05 and 0.02 tn C ac^{-1} y^{-1}) after 10, 30 and 100 years, respectively (Ryals et al. 2015). Further, the slow SOC stock increased by 5.01, 3.72 and 1.15 Mg C ha^{-1} y^{-1} (2.24, 1.66 and 0.51 tn C ac^{-1} y^{-1}) respectively, for the same depth and time intervals. The passive SOC stock increased by 0.02, 0.10 and 0.20 Mg C ha^{-1} y^{-1} (0.01, 0.04 and 0.09 tn C ac^{-1} y^{-1}) after 10, 30 and 100 years, respectively. Overall, total SOC stocks in compost amended simulations remained 6.3% greater than those of the un-amended control simulations after 100 years. However, depending on compost application method and compost C:N ratio, N_2O emissions may also increase and offset some of the climate benefits of SOC sequestration (Ryals et al. 2015).

Another exogenous amendment with potential to improve grassland productivity, and increase soil C inputs and SOC stocks is biochar. C-rich biochar is produced by pyrolysis or thermochemical decomposition of organic material in an oxygen-limited environment under controlled conditions (Lehmann and Joseph 2015). However, biochar effects cannot be generalized as the biological, chemical and physical properties among biochars vary widely. Thus, effects of soil application of biochar on grassland SOC stocks differ not only among biochar types but also among climate, soil types, and land use and management (Lorenz and Lal 2014b). While photosynthesis and biomass may be enhanced by biochar for C3 grassland species, responses of C4 plants may be limited based on the limited global database of long-term field experiments (He et al. 2020). Liu et al. (2016) reported that biochar has a positive effect on grassland SOC but data from field observations on biochar effects on SOC stocks of U.S. grasslands are scanty. Amonette et al. (2021) called for a long-term integrated research program that focuses on biochar. Specifically, there is a need for coordinated, large-scale, regionally focused, long-term studies of biochar production and soil application to answer questions about the technology's cost and potential impact, and to guide future development.

Tundra Biome

Low plant biomass production due to harsh environmental conditions, permafrost and water-saturated conditions contribute to long-term SOC accumulation in tundra soils (Schuur et al. 2018). The dynamic of the SOC stock may be affected by continuous disturbances include increasing temperatures, changing precipitation regimes, and rising atmospheric CO_2 concentrations. Discrete disturbances including wildfire, insect outbreaks, and abrupt permafrost thaw may also affect tundra SOC stocks. Typical tundra fires may be too short-lived to result in substantial soil organic layer consumption but repeated fires may amount to a larger amount of SOC loss than one single severe burning (He et al. 2021). In tundra environments recently affected by fire, soil respiration often exceeds GPP, resulting in a net annual source of CO_2 to the atmosphere (Watts et al. 2021). In contrast, rising summer temperatures are likely stimulating C uptake and storage by plants in tundra areas that are greening, but also leading to SOC loss by thawing permafrost and enhancing microbial decomposition (Berner et al. 2020). However, the majority of Alaskan tundra being an annual ecosystem CO_2 source on average (Virkkala et al. 2021). Thus, the trajectory of the high-latitude SOC stock under climate change is uncertain (Fisher et al. 2014).

The shrub tundra area in Alaska is projected to increase while the graminoid tundra area is projected to decrease (Zhu and McGuire 2016). However, the extent of human management activities to maintain tundra SOC stocks or increase SOC sequestration is rather limited compared to the most productive and accessible portion of the boreal forest biome (Schuur et al. 2018). From 1982 to 2016, land-use activities contributed less than 1% to observed land changes in boreal tundra ecosystems globally (Song et al. 2018). Importantly, there are no proven direct, localized land use management activities that can affect the SOC content of tundra ecosystems (Goldstein et al. 2020).

About 19.5% of Native people in the U.S. live in Alaska (Norris et al. 2012). Many of the native communities in Alaska reside in coastal areas where commercial fishing and tourism are two major sources of income (Tiller 1995). Alaska was not among the top twelve U.S. states of American Indian or Alaska Native agricultural producers in 2017 (USDA 2019). However, there is generally a lack of rigorous studies on land-use impacts to ecosystems on tribal lands including that for tundra. Nevertheless, traditional Indigenous peoples' practices may offer new opportunities for C management (McCarthy et al. 2018).

Warmer temperatures due to climate change and favorable soil properties have recently allowed vegetable production on tundra soils in Alaska (Wyckoff 2015), and it is expected that the tundra biome will be increasingly used for agriculture in the future which will affect SOC sequestration and stocks. For example, Harvey et al. (2021) projected that selecting spring wheat (*Triticum aestivum* L.) varieties will be important to materialize future yield increases in Interior Alaska. Overall, scientists working in the northern regions are cautiously optimistic about climate change–driven expansion of northern agriculture (Altdorff et al. 2021). However, intensification and less adapted practices may be associated with the risk of SOC losses. For example, conversion of northern natural lands to agricultural uses may involve the removal of OM-rich topsoil. Further, when agricultural production includes long term N-fertilization of tundra soils, this may result in decreases in SOC stocks (Mack et al. 2004). Plant N availability in the tundra biome may also be enhanced by the increasing dispersal of earthworms in discontinuous permafrost zones (Blume-Werry et al. 2020). The associated changes in arctic plant communities may affect C cycling and SOC stocks of tundra soils.

3.5.3 Wetland Biomes

Terrestrial Wetlands

Terrestrial wetlands in the U.S. are managed for ecosystem services (e.g., water cycling and filtration; flora and fauna; Kolka et al. 2018), but not for SOC stocks and sequestration (Table 3.1). However, management-induced changes in hydrology, soils or vegetation of wetlands may alter SOC dynamics. Specifically, restoration of drained peatlands to wetlands may result in increases in SOC stocks. Maintaining shallow groundwater levels has the potential to reduce CO_2 emissions (Gewin 2020). However, the biotic and abiotic constraints to wetland restoration are not completely understood, and even after a century of restoration CO_2 sequestration and SOC stocks in restored or created wetlands can be on average lower than those in reference wetlands (Moreno-Mateos et al. 2012).

Rewetted peatlands differ from intact natural peatlands but their ecological functioning is not completely understood (Jurasinski et al. 2020). Thus, the potential to increase SOC stocks by restoration of terrestrial peatlands and wetlands in the U.S. is also uncertain (Chimner et al. 2017). The CH_4 emissions of restored peatlands may increase due to the anaerobic conditions while CO_2 emissions may decrease after restoration (Humpenöder et al. 2020). For example, restored freshwater wetlands in California emitted high amounts of CH_4, turning wetlands into a GHG source while at the same time sequestering C and building peat soil (Hemes et al. 2018). Rewetted peatlands over mid-term have higher GHG emissions than intact peatlands but the GHG mitigation compared to the previous drained status is considerable (Humpenöder et al. 2020). Although the CH_4 cost of rewetting may temporarily be substantial (Hemes et al. 2018), the CO_2 cost of inaction may be much higher. Thus, peatland rewetting can be one important measure to mitigate climate change and attenuate peak global warming (Günther et al. 2020). Overall, restoring degraded peatlands and wetlands in the U.S. appears to have positive effects on GHG mitigation in the long term (Kolka et al. 2018).

For production agriculture in terrestrial wetlands, water levels are controlled to maximize production often at the expense of SOC stocks (Kolka et al. 2018). In the U.S., about 6500 km^2 (2510 mi^2) of peatlands are being used for crop production (ICF International 2013). Crops such as sugarcane (*Saccharum officinarum*), rice (*Oryza sativa* L.), sugar beet (*Beta vulgaris* subsp. *vulgaris*), radish [*Raphanus raphanistrum* subsp. *sativus* (L.) Domin], cranberry (*Vaccinium macrocarpon*), blueberry (*Vaccinium* spp.), lettuce (*Lactuca sativa* L.), celery (*Apium graveolens* L.), carrot [*Daucus carota* subsp. *sativus* (Hoffm.) Schübl. & G. Martens], potato (*Solanum tuberosum* L.), onion (*Allium cepa* L.), and mint (*Mentha* × *piperita* L.) are grown in wetlands but their influence on SOC stocks and C sequestration are often not known. Similarly, aquaculture affects wetlands in the U.S. but little data exist on the impact on C storage or fluxes (Kolka et al. 2018).

Crops grown on peat have among the highest GHG emissions per crop calorie in the world (Carlson et al. 2017). GHG emissions from peatlands drained for agriculture could be greatly reduced without necessarily halting their productive use by adjusting the mean annual effective water table depth, i.e., the average depth of the aerated peat layer (Evans et al. 2021). For example, boreal/temperate peatlands may be net CO_2 sinks when the effective water table depth is <20 cm (8 in) but net CO_2 sources when the effective water table depth is >20 cm (Evans et al. 2021). Current agricultural practices on peat rely on crops derived from dryland ecosystems, which require drainage. Alternatively, cultivating waterlogging tolerant crop varieties may be an option to combine wetland SOC stock maintenance with agricultural production (Verhoeven and Setter 2010). An alternative productive wetland use on shallow water tables is paludiculture or wet agriculture which often competes effectively with drainage-based peatland agriculture with the co-benefit of C sequestration (Joosten et al. 2015). Paludiculture can be defined as sustainable production of biomass on wet and rewetted peatlands (Wichtmann and Joosten 2007). For this management practices, land use, and/or agriculture system, biomass of spontaneous vegetation on natural sites or that of crops and trees established on rewetted sites is produced. Paludiculture trees include alder, and crops include common reed [*Phragmites australis* (Cav.) Trin. ex Steud.], cattails (*Typha* spp.), reed canary grass (*Phalaris arundinacea* L.) and peat moss. However, paludiculture is not widely practiced as technical and political constraints currently hamper large-scale implementation of this productive use of peatland (Joosten et al. 2015). The development of water-tolerant, economically viable crops suitable for cultivation on organic soils should, therefore, be a high priority (Evans et al. 2021). Overall, no data exist on the potential effects of paludiculture in the U.S. on wetland SOC stocks and C sequestration.

The C stock of managed forested peatlands responds to water management and forest harvesting but this may have no long-term effect on the SOC stock (Kolka et al. 2018). The large C stocks of forested peatlands are, however, vulnerable to burning. For example, Alaskan black spruce stands have recently become a net source of C to the atmosphere due to increases in burn area and late-season burning (Turetsky et al. 2011). Fire releases 5 Tg C y^{-1} (5.5 million tn C y^{-1}) from wetlands in Alaska (Lyu et al. 2018). Thus, in the accessible portion of Alaska, management of wildfires in forested peatlands and wetlands may be an opportunity to manage SOC stocks and sequestration. More typical peatland wildfires in the past were low-severity surface burns, and these can increase the stable SOC stock by condensation and aromatization of functional groups, and by physical protection by creating hydrophobic coatings on aggregate surfaces (Flanagan et al. 2020).

Peat utilization as in peat mining for fuel or horticultural purposes is depleting the peatland SOC stock in the U.S. as the peat itself is removed from the wetland. About 426,000 Mg (470,000 tn) of peat are produced annually in CONUS (U.S. Geological Survey 2020). Using substitutes such as (vermi)compost-biochar mixtures instead of peat would be an opportunity to reduce losses and maintain wetland SOC stocks (Alvarez et al. 2017).

3.5.4 Desert and Xeric Shrubland Biome

Many species in deserts are near their physiological limits for temperature and water stress (Julius et al. 2013). Thus, climate change will have profound consequences, in particular, as drought conditions are expected to intensify (Bradford et al. 2020). For example, nonnative grass species may invade, and more frequent wildfires may occur in addition to decreases in soil C, and including carbonate-C stocks (Wilsey et al. 2020). Increased ecosystem C losses may also happen through increase in weathering and erosion (Julius et al. 2013). The southwestern U.S.'s arid and semi-arid drylands exhibited large areas of decrease in short vegetation and large areas of increase in bare ground from 1982 to 2016, indicating long-term land degradation (Song et al. 2018). A nationwide ground survey in the U.S. revealed degradation of soils and vegetation combined with an increased dominance of invasive species in the southwest (Herrick et al. 2010). Climate adaptation resource management plans for restoration of plant communities after disturbance or prolonged degradation need to identify plant species or populations with strong potential to maintain viability for decades to come (Bradford et al. 2020).

Identifying practices that can enhance C sequestration and increase SOC stocks in deserts in the face of changing climate is hampered by often missing accurate baseline data needed to assess management effects (McCarthy et al. 2018). This is particularly the case for deserts managed by Indigenous people. The C fluxes on tribal lands likely differ from those on non-tribal lands as many Indigenous people follow traditional agricultural and land-use practices, and traditional knowledge. Practices on tribal lands may be less intensive, and include: (i) dryland and NT farming; (ii) moving domesticated animals seasonally in accordance with forage availability; (iii) growing legumes and cover crops; and (iv) raising crops and livestock native to ancestral landscapes. The Indigenous communities throughout deserts in the U.S. differ also in their traditional practices (McCarthy et al. 2018). Overall, indigenous practices tend to conserve landscape C stocks. However, the C-carrying capacity of soils of deserts may have been reduced in the past by colonial-driven transformations. For example, Indigenous people have been relocated to reservations on "marginal lands", and forced to convert to Eurocentric crop and livestock production practices. Otherwise, in the arid southwestern U.S., alfalfa potentially contributes to accumulation of SOC under certain temperature and precipitation conditions. Overall, C fluxes for multiple types of tribal agriculture may be close to net neutral (McCarthy et al. 2018).

Some of the deserts in the U.S. are used as grasslands/rangelands with livestock grazing in areas with accessible water (Chapman and Bolen 2015). Overgrazing in those areas may contribute to deterioration of soils, and managing livestock grazing intensity can result increases in SOC stocks but the effects are specific for climate regions and vegetation types. For example, overgrazing in dry lands with low pH and low-clay soils may result in SOC losses (Dlamini et al. 2016). Otherwise, SOC stocks under drier and colder, and under drier and warmer conditions may only increase under low and low to medium grazing intensities, respectively, while those may generally decrease under higher gazing intensities (Abdalla et al. 2018). Establishment of shelterbelts to prevent wind erosion and rotational grazing management could be C sequestration measures on tribal-owned land (McCarthy et al. 2018). Grazing management in some deserts may contribute to increases in SOC stocks but the potential may be small, similar to that of native semiarid rangelands in the U.S. (Sanderson et al. 2020).

3.5.5 *Croplands*

Terrestrial lands of the U.S. have lost 12.2 Pg SOC (13.4 billion tn SOC) to 200-cm (79-in) depth by conversion to croplands (Sanderman et al. 2017). Thus, the cropland area of the U.S. has a large potential to be managed for SOC sequestration and increases in SOC stocks (Table 3.1). For example, Zomer et al. (2017) estimated that available cropland soils (i.e., not high SOC or sandy soils) of the U.S. may sequester 1.2–2.6 Pg SOC (1.3–2.9 billion tn SOC) to 30-cm (12-in) depth annually for at least the 20 years that sequestration can be expected to continue. Increasing SOC concentration (> 2%) contributes also to crop yields in CONUS while being less sensitive to precipitation and temperature (Huang et al. 2021). For example, an increase of 1% SOM in 0–30 cm (0–12 in) depth in all U.S. counties was associated with an average maize yield increase of 0.83 Mg ha^{-1} (0.37 tn ac^{-1}), and of 2.2 Mg ha^{-1} (1.0 tn ac^{-1}) under severe drought, respectively (Kane et al. 2021). Cropland management practices can specifically affect SOC storage by reducing C losses and by increasing C inputs (Paustian et al. 2019). For example, cover-crops, organic fertilizer amendments, moderate grazing or agroforestry may be particularly useful in croplands of the eastern U.S. to increase DOC concentration which potentially contributes to increased sorption and SOC sequestration (Abramoff et al. 2021).

Over the last 70 years, the U.S. yields of soy and winter wheat have more than tripled, while corn yields have increased over five-fold (USDA 2020). Human activity has intensified and amplified the yield geographies explained by sun, soil, and water alone (Burchfield and Nelson 2021). Cropland C inputs in CONUS have also increased between 1971 and 2015 due to farming and biotechnology advances as was indicated by increases in total NPP (Liu et al. 2020a). With adequate water and nutrients, increases in atmospheric CO_2 may have also contributed to increases in crop yields particularly for legumes and root crops (Ainsworth and Long 2021). However, the aim of crop breeding and production is to maximize the amount of (aboveground) harvestable products, and product-C is removed from croplands by harvest while SOC sequestration is currently not focus of farm management (Amelung et al. 2020). For example, about 50% of C is removed as grain from annual grain production systems (Ciampitti et al. 2013; Pedersen and Lauer 2004). Thus, croplands have the lowest percentage and the lowest amount per unit area of total NPP allocated belowground among terrestrial biomes (Gherardi and Sala 2020).

Between 1971 and 2015, the annual C losses by cropland harvest in CONUS were 437 Tg C y^{-1} (482 million tn C y^{-1}; Liu et al. 2020a). Soil erosion does also contribute to C losses from croplands. In 2015, average soil erosion rates on non-federal cropland in the U.S. were estimated at 10.4 Mg ha^{-1} y^{-1} (4.6 tn ac^{-1} y^{-1}; USDA 2018). However, prior estimates of soil degradation from soil survey-based methods may have underestimated soil erosion rates. For example, Thaler et al. (2021) reported that 35% of the cultivated area of the U.S. Corn (*Zea mays* L.) Belt and, in particular, convex hilltop areas have completely lost A-horizon soil with tillage-induced erosion as important driver. Thus, management practices that reduce soil disturbance or increase plant residue input may sequester SOC, and increase SOC stocks in U.S. cropland (Swan et al. 2015). Inherent factors such as precipitation, mean annual temperature and texture in addition to anthropogenic factors also affect SOC across croplands in the U.S. (Nunes et al. 2020a). Specifically, climate change may result in SOC losses from U.S. croplands in the future (Zhang et al. 2020b).

Management

Management practices to increase cropland SOC stocks in the U.S. include: (i) maintaining permanent cropland cover with vegetation (i.e., elimination of summer fallow, use of perennials and cover crops), (ii) protecting the soil from erosion (i.e., reduced tillage or NT, maintaining residue cover), and (iii) improved nutrient and water management (Hristov et al. 2018). From 2017 to 2019, cover crops accounted for an estimated 69,000 km^2 (26,560 mi^2), and NT accounted for 421,000 km^2 (162,500 mi^2) across 1.2 million km^2 (475,000 mi^2) or 90% of farmland of CONUS (Indigo 2020). The USDA (2019) reported a cropland area under NT of 390,427 km^2 (150,744 mi^2) in 2012, and of 422,703 km^2 (163,207 mi^2) in 2017. The overall impacts of NT are more prominent if implemented where it is most needed in vulnerable areas within agricultural watersheds (Lee et al. 2021). Complete conversion of the 0.9 million km^2 (0.3 million mi^2) in corn–soy (*Glycine max* L.) rotation to NT in the U.S. would sequester 21.7 Tg C (23.9 million tn C) annually (Horton et al. 2021).

Removing highly erodible cropland from crop production is another option to reduce losses and increase cropland SOC stocks with an estimated increase by 9.6 Tg C y^{-1} (10.6 million tn C y^{-1}) on 253,000 km^2 (97,700 mi^2; Sperow 2020a).

The SOC stocks may also increase by converting croplands too marginal for crop production, and cropland degraded by long-term cropping into grasslands (Follett et al. 2001). The total marginal land area of CONUS is estimated at 0.55–1.73 million km^2 (0.21–0.67 million mi^2; Yang et al. 2020).

Based on calculations using the COMET-Planer, decreasing fallow frequency and/or adding perennial crops to permanent croplands in the dry/semiarid climate zone has the highest potential for increasing cropland SOC stocks (Table 3.3; Swan et al. 2015). In the moist/humid climate zone, replacing conventional tillage with NT may potentially contribute to increases in SOC stocks. Similarly, NT in dryland cropping systems in arid and semi-arid regions of the U.S. may result in SOC stock increases (Sainju 2020). However, the NT system cannot be used for crops that grow in the soil such as sugarbeet (*Beta vulgaris* subsp. *vulgaris*), potato, sweet potato (*Ipomoea batatas* (L.) Lam.), yam (*Dioscorea* spp.), and carrot (Sainju 2020).

High increases in SOC stocks in the U.S. may also be achieved by converting strips of conventionally managed cropland to permanent unfertilized herbaceous cover (Swan et al. 2015). CA practices that shift to perennials have estimated overall SOC storage rates of 0.42–0.95 Mg C ha^{-1} y^{-1} (0.19–0.42 tn C ac^{-1} y^{-1}) over 20 years, while inclusion of cover crops are estimated to sequester on average 0.15–0.27 Mg C ha^{-1} y^{-1} (0.07–0.12 tn C ac^{-1} y^{-1}) over two decades (Swan et al. 2015).

Sperow (2020a) estimated that >90% of the potential SOC stock increases in topsoil [30-cm (12-in)] depth in U.S. croplands can be realized in moist climate regions, primarily cold and warm temperate regions. The following estimations of cropland SOC stock changes were based on site-specific data on land-use and management, and on factors and initial SOC stocks following IPCC's Tier 1 approach updated 2019 to incorporate a better understanding of SOC changes related to tillage management, grassland management, and land use. Setting aside highly erodible cropland may accumulate 9.6 Tg C y^{-1} (10.6 million tn C y^{-1}) in topsoil over 20 years. Inclusion of winter cover crops may increase total SOC stock by 17.7 Tg C y^{-1} (19.5 million tn C y^{-1}). Reduced bare summer fallow may sequester 3.2 Tg C y^{-1} (3.5 million tn C y^{-1}). Adoption of NT on cropland that was not set-aside was estimated to increase SOC stocks by 18.4 Tg C y^{-1} for at least 20 years (20.3 million tn C y^{-1}, Sperow 2020a). Some uncertainty of the estimated SOC stock changes for croplands in CONUS arise from the use of IPCC factors which may be revised in the future. Further, enhanced residue inputs from plant genetic improvements, better fertilizer management or improved pest control were not considered. Also, it was assumed that SOC stocks reach a steady state or equilibrium after 20 years while inclusion of cover crops may not result in steady state SOC stocks after more than a century (Poeplau and Don 2015). Adoption of land-use and management activities that increase SOC stocks depends also upon the effect of potential change in crop yields and input costs on profit (Sperow 2020a). For example, 95% of the biophysical potential SOC sequestration increase on U.S. cropland by NT adoption could be captured for less than \$100 Mg^{-1} CO_2 (\$91 tn^{-1} CO_2; Sperow 2020b).

Table 3.3 Average soil organic carbon sequestration rates over 20 years under United States Department of Agriculture National Resources Conservation Service conservation practices for croplands including row crops, close-grown crops, hay or pasture in rotation with cultivated crops, continuous hay, perennial crops, horticultural cropland, land with agroforestry dominantly used for crop production, and lands in temporary fallow or enrolled in conservation reserve programs (negative values indicate losses; modified from Swan et al. 2015; U.S. EPA 2021)

NRCS Conservation Practice	Climate zone	Soil organic carbon sequestration rate	
		Mg C ha^{-1} y^{-1}	tn C ac^{-1} y^{-1}
Conventional tillage to no-till	Dry/semiarid	0.01 to 0.33	0.01 to 0.15
	Moist/humid	0.08 to 0.47	0.04 to 0.21
Conventional tillage to reduced tillage	Dry/semiarid	0.02 to 0.12	0.01 to 0.05
	Moist/humid	0.01 to 0.13	0.01 to 0.06
Replacing synthetic nitrogen fertilizer with soil amendments	Dry/semiarid	0.24 to 1.33[a]	0.11 to 0.59[a]
	Moist/humid	0.52 to 1.53[a]	0.23 to 0.68[a]
Conservation crop rotation: Decreasing fallow frequency and/or adding perennial crops	Dry/semiarid	−0.11 to 0.43	−0.05 to 0.19
	Moist/humid	0.00 to 0.30	0.00 to 0.13
Cover crops: Addition of seasonal cover crops to annual cropland	Dry/semiarid	0.05 to 0.27	0.02 to 0.12
	Moist/humid	0.10 to 0.28	0.04 to 0.13
Stripcropping: Growing perennial cover in strips with annual crops	Dry/semiarid	0.00 to 0.15	0.00 to 0.07
	Moist/humid	0.00 to 0.15	0.00 to 0.07
Mulching	Dry/semiarid	0.05 to 0.27[a]	0.02 to 0.12[a]
	Moist/humid	0.10 to 0.28[a]	0.04 to 0.13[a]
Conservation cover: Conventionally managed cropland to permanent unfertilized herbaceous cover	Dry/semiarid	0.42 to 0.86	0.19 to 0.38
	Moist/humid	0.39 to 0.82	0.17 to 0.37
Forage and Biomass Planting: Full conversion – replacing all crops in a conventionally managed continuous grain rotation with continuous unfertilized forage/biomass crops	Dry/semiarid/moist/humid	−0.21 to 0.49	−0.10 to 0.22

(continued)

Table 3.3 (continued)

NRCS Conservation Practice	Climate zone	Soil organic carbon sequestration rate	
		Mg C ha^{-1} y^{-1}	tn C ac^{-1} y^{-1}
Forage and Biomass Planting: Partial conversion – replacing a conventionally managed continuous grain rotation with a grain – unfertilized forage/biomass crop rotation	Dry/semiarid/moist/humid	0.00 to 0.30	0.00 to 0.13
Herbaceous Wind Barriers/Vegetative Barriers/Riparian Herbaceous Cover/Contour Buffer Strips/Field Border/Filter Strip/Grassed Waterway: Converting strips of conventionally managed cropland to permanent unfertilized herbaceous cover	Dry/semiarid	0.42 to 0.86	0.19 to 0.38
	Moist/humid	0.39 to 0.82	0.17 to 0.37
Windbreak/Shelterbelt Establishment: Replacing a strip of conventionally managed and fertilized cropland with unfertilized woody plants	Dry/semiarid	0.23 to 1.79[b]	0.10 to 0.80[b]
	Moist/humid	0.54 to 2.20[b]	0.24 to 0.98[b]
Windbreak/Shelterbelt Renovation: Replacing half of woody plants in an existing unfertilized windbreak	Dry/semiarid	0.04 to 0.61[c]	0.02 to 0.27[c]
	Moist/humid	0.07 to 0.66[c]	0.03 to 0.29[c]
Hedgerow Planting: Replacing conventionally managed and fertilized cropland with unfertilized woody plants	Dry/semiarid	0.29 to 0.71[b]	0.13 to 0.32[b]
	Moist/humid	0.65 to 1.15[b]	0.29 to 0.51[b]
Alley Cropping/Multistory Cropping: Replacing 20% of the area of conventionally managed and fertilized cropland with woody plants	Dry/semiarid	0.00 to 0.99[b]	0.00 to 0.44[b]
	Moist/humid	0.00 to 2.10[b]	0.00 to 0.93[b]

[a]Including soil organic carbon stock increases due to external soil carbon inputs
[b]Including woody biomass carbon accumulation
[c]Including new woody biomass carbon accumulation

Conservation Agriculture Practices

Implementation of CA practices in U.S. croplands may result in increases in SOC stocks (Table 3.3). The CA practices include: (i) continuous residue cover on the soil surface, (ii) continuous minimum soil disturbance (NT), and (iii) diverse crop rotations and cover crop mixes (Reicosky 2015). The potential for SOC storage under CA may further increase by extending soil lifespans and promoting soil thickening (Evans et al. 2020). However, effects of CA practices (i.e., NT plus residue retention) on cropland SOC stocks in the U.S. are variable (Page et al. 2020). For example, decreases by 0.15 Mg C ha^{-1} y^{-1} (0.07 tn C ac^{-1} y^{-1}) to 60-cm (24-in) depth in the Midwest, and increases by 0.45 Mg C ha^{-1} y^{-1} (0.20 tn C ac^{-1} y^{-1}) in 15–30 cm depth (6–12 in) in the Southeast have been reported. Thus, the effects of CA on cropland SOC stocks in the U.S. are highly site-specific (Page et al. 2020).

Cover Crops

Previous trends in the U.S. estimated cover crop adoption across 81,000 km^2 (31,000 mi^2) likely in the year 2020, over which cover crops were estimated to have sequestered 16.3 Tg C y^{-1} (18.0 million tn C y^{-1}) in soil (Table 3.3; Tellatin and Myers 2018). Also, cover crops prevent soil erosion, thus retaining SOC and preventing emissions associated with the introduction of C into aquatic ecosystems with emissions estimated at 0.78 Tg CO_2e y^{-1} (0.86 million tn CO_2e y^{-1}). Schlautman et al. (2021) proposed integrating perennial groundcovers (living mulches or perennial cover crops) into annual cash-crop systems in the U.S. for improving SOC storage. However, any herbaceous cover on cropland should not be used as grazed pasture for livestock as associated GHG emissions may negate the abatement from increased SOC stocks (Meier et al. 2020). Specifically, there may be a trade-off in SOC stock gains when using cover crops for grazing versus that for use as green manure (Sarkar et al. 2020). Also, cover crops may have only small effects on crop yield. For example, Seifert et al. (2018) estimated average yield increases of 0.65% for maize and 0.35% for soybean in the U.S. Midwest in 2008–2016.

Tillage

Adoption of NT has been slow in the U.S. primarily due to social and economic reasons (Sainju 2020). NT management has also become a controversial approach for storing SOC due to conflicting findings (Table 3.3; Ogle et al. 2019). The highest SOC sequestration per hectare among global regions by adopting NT was simulated for temperate regions of North America (Graham et al. 2021). However, there are large uncertainties in field estimates of SOC sequestration under NT reported in the literature. Globally, the capacity of NT practices to offset current GHG emissions through SOC sequestration may be more limited than has been previously anticipated (Graham et al. 2021). Often, the depth distribution of SOC is altered under NT management with no overall change in soil profile SOC stocks. More field data with deeper sampling depths and longer duration are essential for a more credible assessment of effects of NT practices on SOC stocks (Xiao et al. 2020). For example, Nicoloso and Rice (2021) reported based on a global meta-analysis that NT soils stored 6.7 Mg SOC ha^{-1} (3.0 tn SOC ac^{-1}) more than tilled soils (0–100 cm [0–39 in] depth) with an average of 16 years of NT, in contrast to previous findings.

The impact of NT on cropland biological soil health indicators in the U.S. is also affected by latitude, soil order, time under NT, and cropping system (Nunes et al. 2020c). For example, the positive response of topsoil SOC (0 to $\leq$15 cm [0 to $\leq$6 in]) to NT system compared to moldboard plow (MP) was significant only for Alfisols, Inceptisols, Mollisols, and Ultisols (Nunes et al. 2020c). NT affected also chemical soil health indicators in U.S. cropping systems (Nunes et al. 2020b). For example, topsoil total N response to NT was moderated by soil order and cropping system, with the largest increase in total N found in Ultisols, Inceptisols, Alfisols, and Mollisols under more diversified cropping systems including those with cover crops. Further, the greatest topsoil P increase in response to NT was found under long-term experiments (>5-yr) and on fine-textured soils. Compared to MP, NT increased total N, P, and K concentrations within the top 15 cm (6 in). Below that depth, Ca and Mg concentrations were lower under NT than MP but total N, P, and K were not different (Nunes et al. 2020b).

Subsoiling is a tillage management practice with the purpose to loosen soil structure, thereby decreasing subsoil strength without disturbing soil horizons (Feng et al. 2020). Subsoiling can increase macroporosity in soil, which enhances root extension and plant development and in turn increases root-derived C input. Results from a global meta-analysis indicated that subsoiling increased SOC stock by 11.6, 9.6, 6.2 and 9.3% in 0–10, 10–20, 20–30 and 50–100 cm (0–3.9, 3.9–7.9, 7.9–11.8 and 19.7–39.4 in) depth, respectively, compared to conventional tillage (Feng et al. 2020). The individual response of SOC stock to subsoiling was found to be highly site specific with greater benefits in soil under arid zones and under rotational cropping systems. Thus, SOC stocks of some croplands in the U.S. may be increased by subsoiling but data are scanty.

Fertilization

The application of commercial fertilizers may increase crop yields in the U.S. by 40–60% (Stewart et al. 2005). The associated increases in crop-derived soil C inputs may contribute to increases in cropland SOC stocks. Specifically, fertilizer-induced enhancement of formation of deep and extended crop root systems, and increased root turnover may result in increases in SOC stock (Rasse et al. 2005). For example, belowground biomass of maize (*Zea mays* L.) more strongly contributes to stabilized SOC stocks over the medium to long term compared to aboveground residues (Xu et al. 2021a). Similar to grassland soils, N additions to cropland soils that previously received little or no additional N could contribute to SOC sequestration (Huang et al. 2020). In contrast, further N additions to cropland systems that already receive substantial amounts of fertilizer N are unlikely to stimulate new SOC storage.

Biochar

A soil amendment with potential agronomic benefits by mitigating yield-limiting soil properties in U.S. croplands may be biochar. However, knowledge must be improved as many studies have focused on highly weathered, nutrient-poor, or acidic soils, for example, agricultural soils from the southeastern U.S. Coastal Plain (Zheng et al. 2015). For the better quality soils such as those from the Midwestern U.S., a synergistic agronomic effect has only been observed when the use of biochar was combined with chemical fertilizer. Average increases in crop yields may range from 4.7% to 6.4% depending on biochar feedstock and application rates (Dokoohaki et al. 2019). Largest biochar-induced yield increases were predicted in the U.S. Southeast and the Northeast which are also regions that have limited cropland and low quality soils (Dumortier et al. 2020). Applying biochar to cropland under corn may be most profitable from a revenue perspective compared to soybeans and wheat because additional revenues accrued by farmers may be not enough to cover the costs of biochar applications in many U.S. regions. Economic analyses suggest that deployment of a biochar/bioenergy platform would be most successful in the Southeast region of the U.S. where woody biomass is readily available as feedstock (Dumortier et al. 2020). Further, photosynthesis and biomass of C3 plants such as soybeans and wheat may benefit more from biochar than that of C4 plants, e.g., corn (He et al. 2020). Otherwise, adding C-rich biochar to soil may also directly increase the SOC stock (Han et al. 2020), but whether adding biochar produced from atmospheric CO_2 fixed outside the cropland can be regarded as SOC sequestration in the cropland is a debatable issue. An integrated long-term research program on biochar is needed (Amonette et al. 2021).

Irrigation

Some U.S. cropland area, especially in arid and semi-arid regions is managed intensively by irrigation to reduce soil moisture limitations on crop production (Hristov et al. 2018). In 2015, about 200,000 km^2 (77,000 mi^2) cultivated cropland were irrigated while 60,000 km^2 (23,000 mi^2) of non-cultivated cropland received irrigation (USDA 2018). Cultivated cropland comprises land in row crops or close-grown crops, and also other cultivated cropland, for example, hayland or pastureland that is in a rotation with row or close-grown crops. Non-cultivated cropland includes also horticultural cropland (USDA 2018). In 1950 to 2016, irrigation resulted in increases of yields for spring wheat (147%), corn silage (132%), corn grain (124%), barley (*Hordeum vulgare* L.; 93%), winter wheat (79%), soybean (72%), sorghum (*Sorghum bicolor* (L.) Moench; 70%), and cotton (*Gossypium arboretum* L.; 64%) compared to non-irrigated yields in CONUS (Lu et al. 2020). Soil C inputs may have also increased with irrigation-induced crop yield increases.

The irrigation of cropped soils in arid and semiarid regions in the U.S. may lead to increases in SOC stocks (Trost et al. 2013). For example, after 30–40 years SOC stocks in 0–20 cm (0–8 in) soil depth of a corn/wheat/soybean rotation in an arid region in Colorado was 22 Mg C ha^{-1} (10 tn C ac^{-1}) and 30 Mg C ha^{-1} (13 tn C ac^{-1}) under non-irrigated and irrigated management, respectively (Denef et al. 2008). In contrast, no consistent irrigation effects on cropland SOC stocks were observed in humid regions but data from long-term field experiments monitoring irrigation effects on SOC stocks in the U.S. are scanty. The net effects of irrigation on cropland GHG fluxes and GWP, particularly, for other crops than rice (*Oryza sativa* L.) must also be addressed (Sapkota et al. 2020).

Whether high crop yields under irrigation can be sustained in the future is a debatable issue. For example, the strong yield signal in the High Plains Aquifer highlights production intensification brought by heavy groundwater irrigation (Burchfield and Nelson 2021). However, this aquifer is expected to lose an estimated 24% of irrigated area in this century due to agricultural withdrawals (Deines et al. 2019). How this loss will affect cropland SOC stocks is uncertain.

Crop Rotation

Crop species within a rotation may be selected for increased soil C inputs and associated increases in SOC stocks (Table 3.3). Specifically, high crop NPP may contribute to increases in SOC stocks with simulated corn NPP in the U.S. Corn Belt in the range of 7–11 Mg C ha^{-1} (3–5 tn C ac^{-1}; Zhang et al. 2020a). In comparison, NPP of soybeans was much lower in the range of 3–5 Mg C ha^{-1} (1–2 tn C ac^{-1}) in most counties. Comparable values for NPP were predicted for winter wheat in the Midwest (Zhang et al. 2020a). Globally, maize and ryegrass (*Lolium perenne* L.) had the greatest C allocation to the soil (1.0 Mg C ha^{-1} y^{-1} [0.5 tn C ac^{-1} y^{-1}]), followed by wheat (0.8 Mg C ha^{-1} y^{-1} [0.4 tn C ac^{-1} y^{-1}]), and rice (0.7 Mg C ha^{-1} y^{-1} [0.3 tn C ac^{-1} y^{-1}]) among 227 observations from different experiments (Mathew et al. 2020). Mwafulirwa et al. (2021) suggested that maize varieties with decreased morphological trait root diameter but increased root length decreased SOC mineralization, and genotypic variation in traits should be exploited within breeding programs. However, exactly which root characteristics and traits are important for maximizing SOC gains and for ensuring long-term carbon storage is not obvious (Jansson et al. 2021). Thus, whether cropland SOC stocks in the U.S. increase by including crop species with higher NPP and C allocation to the soil more frequently in a rotation compared to crops with lower NPP and C allocation to the soil needs to be studied. Similarly, whether increases in plant species mixtures in croplands results in increases in SOC stock is a researchable priority (Chen et al. 2020a).

Aside the adoption of new technologies and management practices, crop adaptation may be needed to increase the resistance of U.S. cropland SOC stocks to loss by climate change. For example, Zhang et al. (2020b) predicted that changing planting dates and varieties in the U.S. Corn Belt may result in no differences in SOC stocks to 20-cm (8 in) depth under different climate change scenarios. In contrast, climate change would lead to SOC losses in this region without adaptation (Zhang et al. 2020b).

Intercropping, i.e., the mixed cultivation of crop species on the same field, is currently not a part of high-input and high-yield cropland agriculture in the U.S. (Li et al. 2020). However, intercropping that includes maize may increase the already high cereal yields by 20–30% (Tilman 2020), which may translate into increases in SOC stocks.

Perennials

Increasing the proportion of perennials in a crop rotation (e.g., grasslands, pasture, hayed lands, tree crops, and vineyards) may result in increases in cropland SOC stocks in the U.S. (Table 3.3; Hristov et al. 2018). For example, a sod-based crop rotation system did sequester or at least maintain SOC stocks comparable to that of a native ecosystem in 0–30 cm (0–12 in) depth under sustainable management practices in the Southeastern Plains (Rolando et al. 2021). The grassland duration should probably exceed the duration of crop phases in a rotation to increase SOC stocks (Crème et al. 2020). However, the magnitude of SOC gain by including perennials in a rotation may be strongly dependent on the site history, and on how far the SOC stock is away from equilibrium. In contrast, a substantial reduction in SOC stock may nevertheless occur on perennial forage cropped soils. For example, including alfalfa in a rotation in Minnesota reduced C losses by 23% relative to continuous silage maize, but net losses of −3.8 Mg C ha^{-1} y^{-1} (−1.7 tn C ac^{-1} y^{-1}) were still observed (Gamble et al. 2021).

Increased integration of trees in croplands may particularly accumulate SOC aside that of biomass C (Chapman et al. 2020). Perennial crops can be defined as crops that are planted, but not replanted and/or fully harvested annually to obtain goods (Ledo et al. 2020). Tillage and soil disturbances are reduced in perennial crop cultivation. Perennial crops and grasses have longer growing and photosynthetic active seasons, increased seasonal light interception efficiencies, deeper rooting depths, and greater root mass compared to annual crops (Glover et al. 2010b). Greater soil C inputs by increased biomass produced and left on site together with reduced erosion risk may contribute to higher SOC stocks under perennial crops and grasses. For example, perennial grass fields in Kansas had 43 Mg C ha^{-1} (19 tn C ac^{-1}) higher SOC stocks to 1-m (39-in) depth compared to annual crop fields (Glover et al. 2010a).

Ledo et al. (2020) reported, based on a global meta-analysis, that a change from annual to perennial crops induces a SOC stock gain. Specifically, a change from annual to perennial crops may lead to an average 20% increase in SOC stock at 0–30 cm (0–12 in) (6.0 Mg C ha^{-1} [2.7 tn C ac^{-1}]), and a total 10% increase in 0–100 cm (0–39 in) soil depth (5.7 Mg C ha^{-1} [2.5 tn C ac^{-1}]) over a 20-year period or longer. The potential SOC accumulation rates in fields converted from annual to perennial grains have been estimated to range globally between 0.13 and 1.70 Mg C ha^{-1} y^{-1} (0.06 and 0.76 tn C ac^{-1} y^{-1}; Crews and Rumsey 2017). Thus, perennial crops may also be included in crop rotations to sequester SOC in U.S. croplands (Ledo et al. 2020).

Perennial vegetables may also play a role in the perennialization of U.S. croplands to increase SOC sequestration (Toensmeier et al. 2020). Perennial vegetables live for 3 or more years, and can be categorized by form into woody plants, vines, and herbs. Woody plants include trees, shrubs, bamboos (*Bambusoideae*), palms (*Arecaceae*), cacti (*Cactaceae*), and woody succulents. Perennial vines include both lianas and herbaceous vines. Perennial herbs include forbs, ferns (*Polypodiopsida* or *Polypodiophyta*), grasses and grass-like plants. Tree vegetables, in particular, feature high C sequestration rates. To realize the full potential of perennial vegetables, more research is needed on this neglected and underutilized class of crops (Toensmeier et al. 2020).

Agroforestry

Among the land-use systems on U.S. croplands including perennials is agroforestry which may potentially result in SOC stock gains (Table 3.3). Agroforestry refers to the production of crop, livestock, and tree biomass on the same area of land (Lorenz and Lal 2014a). Globally, homegarden and silvopastoral systems may have the most considerable potential for SOC sequestration to 30-cm (12-in) depth (Menichetti et al. 2020). Common agroforestry practices with crops in the U.S. include alley cropping (trees or shrubs planted in sets of single or multiple rows with agronomic or horticultural crops produced in the alleys between the trees that can also produce additional products), and windbreaks (single or multiple rows of trees or shrubs that are established for environmental purposes, and may be referred to as crop or field windbreak, livestock windbreak, living snow fence, farmstead windbreak, or hedgerow; Schoeneberger et al. 2017). Producers use windbreaks on agricultural lands in the U.S. mostly for soil erosion control (Smith et al. 2021).

Simultaneously growing crops and trees on the same piece of land (agrisilviculture) may sequester SOC in North America at a rate of 0.39 Mg C ha^{-1} y^{-1} (0.17 tn C ac^{-1} y^{-1}) in 0–30 cm (0–12 in) depth (Feliciano et al. 2018). However, the direction and magnitude of change in SOC stock in the U.S. will depend on the ecological context of the site and the type of agroforestry system implemented (Schoeneberger et al. 2017). Inherently highly variable SOC stocks have been found in agroforestry systems compared with nearby forest-only plantation and treeless croplands. This may explain the variability of agroforestry effects on SOC stocks reported. Methodological difficulties, including differences in sampling depth and selection of the site to provide comparative baselines, further limit assessments to qualitative rather than more quantitative comparisons. Thus, results to date from agroforestry studies in the U.S. indicate SOC sequestration under agroforestry may be negligible/undetectable to possibly negative for several years after initial establishment. However, compared to croplands, agroforestry systems tend to store significantly more SOC deeper in the soil profile, and in the smaller sized fractions, all of which contribute a greater stability to the sequestered SOC (Schoeneberger et al. 2017).

Cropland complexity

Cropland expansion and intensification have simplified the landscape of cropland, thereby adversely affecting the biodiversity and ecosystem services that support agricultural production. Based on data for counties in CONUS from 2008 to 2018, Nelson and Burchfield (2021) reported that increased landcover diversity was associated with yield increases for corn and wheat of more than 10% – an effect strength similar to the impact of seasonal precipitation and soil suitability. Further, landscape configurations that were both moderately complex and also highly diverse were associated with yield increases of more than 20% for corn and wheat (Nelson and Burchfield 2021). Thus, increasing the complexity of landcover of croplands may provide a way to improve crop productivity in the U.S. which may also result in increases in cropland SOC stocks.

3.5.6 Urban Areas

The SOC stock of settlements remaining settlements in the U.S. decreased in the year 2019 (Table 3.1; U.S. EPA 2021). Specifically, urban SOC stocks are heavily affected by human activities, especially those associated with land clearing, removal of vegetation, and disturbance of soils associated with construction activities (Gurney et al. 2018). Specifically, urban SOC stocks may differ by effects of urban land use and management activities including those associated with additions of natural and technogenic materials, and physical soil disturbance by excavation, export and mixing, and soil sealing (Lorenz and Lal 2009, 2017). Soil processes respond to similar human management and resource inputs across urban areas in the U.S. (Trammell et al. 2020). However, management of urban SOC may be an overlooked opportunity to enhance SOC stocks and improve C budgeting of terrestrial biomes (Scharenbroch et al. 2017).

Carbon management decisions in urban areas do not generally consider effects on SOC stocks as, for example, open soils in urban areas are not managed for their SOC stocks but for supporting green infrastructure including urban agriculture (Lorenz 2017). Another objective of urban soil management is water infiltration to offset losses in infiltration capacity by soil sealing under built infrastructure. However, SOC stocks differ among urban land uses. For example, SOC stocks to 1-m (39 in) depth were greater in the most urbanized land-use (commercial, industrial, transportation, utility and vacant) (450 Mg C ha^{-1} [201 tn C ac^{-1}]) compared to forest, park, residential and agricultural (365, 366, 365 and 246 Mg C ha^{-1} [163, 163, 163 and 110 tn C ac^{-1}], respectively) lands across the Chicago region (Scharenbroch et al. 2017). Anthropogenic factors had the most influence on urban SOC stocks after soil properties. An assessment of SOC stocks in New York City, as part of an USDA-NRCS soil survey, indicated that SOC stocks to 1-m (39-in) depth of urban soils under woodland cover were 137 Mg C ha^{-1} (61 tn C ac^{-1}), and 154 Mg C ha^{-1} (69 tn C ac^{-1}) under turfgrass cover (Lorenz and Shaw 2017). Similar assessments are not available for many other urban areas across the U.S., and any recommendations on soil and land use management practices to increase SOC stocks in urban areas of the U.S. would be premature.

Urban soils receiving substantial organic C inputs by management and those characterized by reduced decomposition rates may have increased SOC stocks (Trammell et al. 2020). Management practices such as addition/burial of organic C (e.g., manure, plant litter, construction debris), fertilization and irrigation potentially increase SOC stocks in urban soils in the U.S. (Lorenz and Lal 2015). Among site-specific practices are revegetation of bare urban soils, and fertilizing, irrigation, reduced soil disturbance and residue management (i.e., returning grass clippings) in green space soils. During construction activities, urban soils can potentially be improved toward SOC accumulation by adding organic amendments such as biosolids, and yard and food wastes. Specifically, urban grassland, lawn or turfgrass SOC stocks may benefit from management practices as those lands are often intensively managed by fertilization, irrigation and mowing. However, GHG emissions from management practices may decrease or completely offset the climate benefit of urban grassland SOC stocks (Townsend-Small and Czimczik 2010).

The potential total SOC sink capacity of home lawns in the U.S. was estimated at 496 Tg C (547 million tn C), leading to a C-positive land use for between 66 and 199 years after considering hidden C costs of mowing and fertilization (Selhorst and Lal 2013). Further, diversifying urban grasslands may be a low-risk opportunity to increase SOC stocks in developed landscapes (Thompson and Kao-Kniffin 2017). Garden management practices such as the addition of peat, composts and mulches, and cultivation of trees and shrubs, may also contribute to greater SOC stocks in urban gardens (Lorenz and Lal 2015).

3.6 Conclusions

Managing terrestrial biomes in the U.S. for SOC sequestration is a relatively new objective with benefits for the climate, soil health and quality, the environment and society. SOC sequestration for climate change adaptation and mitigation occurs when the protected and stabilized SOC stock increases, and when soil C inputs are derived from atmospheric CO_2 captured within the biome area. This process continues as long as soils are in disequilibrium – until a new balance between carbon inputs and outputs is reached. The accessible and manageable portions of forest biomes, and of crop- and grasslands can be particularly managed for increasing plant-derived soil C inputs, and for reducing SOC losses by a range of soil and land-use management practices. The effects of climate change and, more generally, global change on SOC sequestration in terrestrial biomes of the U.S. are, however, uncertain. Major reasons of concern are the recently altered fire regimes, particularly in forest biomes of the western U.S., and in the boreal forest/taiga biome in Alaska were large SOC stocks are currently sequestered in peatlands and permafrost.

References

Abdalla M, Hastings A, Chadwick DR et al (2018) Critical review of the impacts of grazing intensity on soil organic carbon storage and other soil quality indicators in extensively managed grasslands. Agric Ecosyt Environ 253:62–81. https://doi.org/10.1016/j.agee.2017.10.023

Abramoff RZ, Georgiou K, Guenet B et al (2021) How much carbon can be added to soil by sorption? Biogeochemistry 152:127–142. https://doi.org/10.1007/s10533-021-00759-x

Achat DL, Fortin M, Landmann G, Ringeval B, Augusto L (2015) Forest soil carbon is threatened by intensive biomass harvesting. Sci Rep 5:15991. https://doi.org/10.1038/srep15991

Adams MB, Kelly C, Kabrick J, Schuler J (2019) Temperate forests and soils. In: Busse M, Giardina CP, Morris DM, Page-Dumroese DS (eds) Global change and forest soils, Developments in soil science 36. Elsevier, Amsterdam, pp 83–108. https://doi.org/10.1016/B978-0-444-63998-1.00006-9

Adewopo JB, Silveira ML, Xu S et al (2014) Management intensification impacts on soil and ecosystem carbon stocks in subtropical grasslands. Soil Sci Soc Am J 78:977–986. https://doi.org/10.2136/sssaj2013.12.0523

Adhikari K, Owens PR, Libohov Z et al (2019) Assessing soil organic carbon stock of Wisconsin, USA and its fate under future land use and climate change. Sci Total Environ 667:833–845. https://doi.org/10.1016/j.scitotenv.2019.02.420

Ainsworth EA, Long SP (2021) 30 years of free-air carbon dioxide enrichment (FACE): what have we learned about future crop productivity and its potential for adaptation? Glob Chang Biol 27:27–49. https://doi.org/10.1111/gcb.15375

Albaugh TJ, Fox TR, Cook RL et al (2018) Forest fertilizer applications in the southeastern United States from 1969 to 2016. For Sci 65:355–362. https://doi.org/10.1093/forsci/fxy058

Alizadeh MR, Abatzoglou JT, Luce CH et al (2021) Warming enabled upslope advance in western US forest fires. Proc Natl Acad Sci USA 118:e2009717118. https://doi.org/10.1073/pnas.2009717118

Altdorff D, Borchard N, Young EH et al (2021) Agriculture in boreal and Arctic regions requires an integrated global approach for research and policy. Agron Sustain Dev 41:23. https://doi.org/10.1007/s13593-021-00676-1

Alvarez JM, Pasian C, Lal R, López R, Fernández M (2017) Vermicompost and biochar as substitutes of growing media in ornamental-plant production. J Appl Hortic 19:205–214

Amelung W, Bossio D, de Vries W et al (2020) Towards a global-scale soil climate mitigation strategy. Nat Commun 11:5427. https://doi.org/10.1038/s41467-020-18887-7

Amiro BD, Barr AG, Barr JG et al (2010) Ecosystem carbon dioxide fluxes after disturbance in forests of North America. J Geophys Res 115:G00K02. https://doi.org/10.1029/2010JG001390

Amonette JE, Blanco-Canqui H, Hassebrook C et al (2021) Integrated biochar research: a roadmap. J Soil Water Conserv 76:24A–29A. https://doi.org/10.2489/jswc.2021.1115A

Anderson RC (2006) Evolution and origin of the Central Grassland of North America: climate, fire, and mammalian grazers. J Torrey Bot Soc 133:626–347

Anderson-Teixeira KJ, Herrmann V, Morgan RB et al (2021) Carbon cycling in mature and regrowth forests globally. Environ Res Lett 16:053009. https://doi.org/10.1088/1748-9326/abed01

Andrieux B, Beguin J, Bergeron Y, Grondin P, Paré D (2018) Drivers of post-fire organic carbon accumulation in the boreal forest. Glob Chang Biol 24:4797–4815. https://doi.org/10.1111/gcb.14365

Angst G, Mueller KE, Eissenstat DM et al (2019) Soil organic carbon stability in forests: distinct effects of tree species identity and traits. Glob Chang Biol 25:1529–1546. https://doi.org/10.1111/gcb.14548

Bailey DW, Mosley JC, Estell RE et al (2019) Synthesis paper: targeted livestock grazing: prescription for healthy rangelands. Rangel Ecol Manag 72:865–877. https://doi.org/10.1016/j.rama.2019.06.003

Balch JK, Bradley BA, Abatzoglou JT et al (2017) Human-started wildfires expand the fire niche across the United States. Proc Natl Acad Sci USA 114:2946–2951. https://doi.org/10.1073/pnas.1617394114

Ballantyne AP, Liu Z, Anderegg WRL et al (2021) Reconciling carbon-cycle processes from ecosystem to global scales. Front Ecol Environ 19:57–65. https://doi.org/10.1002/fee.2296

Beamesderfer ER, Arain MA, Khomik M, Brodeur JJ, Burns BM (2020) Response of carbon and water fluxes to meteorological and phenological variability in two eastern North American forests of similar age but contrasting species composition – a multiyear comparison. Biogeosciences 17:3563–3587. https://doi.org/10.5194/bg-17-3563-2020

Berner LT, Massey R, Jantz P et al (2020) Summer warming explains widespread but not uniform greening in the Arctic tundra biome. Nat Commun 11:4621. https://doi.org/10.1038/s41467-020-18479-5

Bernier P, Paré D, Stinson G et al (2017) Moving beyond the concept of "primary forest" as a metric of forest environment quality. Ecol Appl 27:349–354. https://doi.org/10.1002/eap.1477

Berryman E, Hatten J, Page-Dumroese DS et al (2020) Soil carbon. In: Pouyat R, Page-Dumroese D, Patel-Weynand T, Geiser L (eds) Forest and rangeland soils of the United States under changing conditions. Springer, Cham, pp 9–31. https://doi.org/10.1007/978-3-030-45216-2_2

Bhatti J, Jassal R, Black TA (2012) Decarbonization of the atmosphere: role of the boreal forest under changing climate. In: Lal R, Lorenz K, Hüttl RF, Schneider BU, von Braun J (eds) Recarbonization of the biosphere. Springer, Dordrecht, pp 203–228. https://doi.org/10.1007/978-94-007-4159-1_10

Bigelow DP, Borchers A (2017) Major uses of land in the United States, 2012. EIB-178, U.S. Department of Agriculture, Economic Research Service

Binkley D, Fisher RF (2020) Ecology and management of forest soils, 5th edn. Wiley, New York

Blanco JA (2018) Managing forest soils for carbon sequestration: insights from modeling forests around the globe. In: Muñoz MÁ, Zornoza R (eds) Soil management and climate change.

Academic, Cambridge, MA, pp 237–252. https://doi.org/10.1016/B978-0-12-812128-3.00016-1

Blanco-Canqui H (2021) No-till technology has limited potential to store carbon: how can we enhance such potential? Agric Ecosyst Environ 313:107352. https://doi.org/10.1016/j.agee.2021.107352

Blume-Werry G, Krab EJ, Olofsson J et al (2020) Invasive earthworms unlock arctic plant nitrogen limitation. Nat Commun 11:1766. https://doi.org/10.1038/s41467-020-15568-3

Blunden J, Boyer T (eds) (2020) State of the climate in 2020. Bull Am Meteor Soc 102:Si–S475. https://doi.org/10.1175/2021BAMSStateoftheClimate.1

Boča A, van Miegrot H, Gruselle MC (2014) Forest overstory effect on soil organic carbon storage: a meta-analysis. Soil Sci Soc Am J 78:S35–S47. https://doi.org/10.2136/sssaj2013.08.0332nafsc

Boose E, Foster DR, Fluet M (1994) Hurricane impacts to tropical and temperate forest landscapes. Ecol Monogr 64:369–400

Bormann BT, Spaltenstein H, McClellan MH et al (1995) Rapid soil development after windthrow disturbance in pristine forests. J Ecol 83:747–757

Bormann BT, Homann PS, Darbyshire RL, Morrissette BA (2008) Intense forest wildfire sharply reduces mineral soil C and N: the first direct evidence. Can J For Res 38:2771–2783. https://doi.org/10.1139/X08-136

Bossio DA, Cook-Patton SC, Ellis PW et al (2020) The role of soil carbon in natural climate solutions. Nat Sustain. https://doi.org/10.1038/s41893-020-0491-z

Bouchard M, Pothier D, Ruel JC (2009) Stand-replacing windthrow in the boreal forests of eastern Quebec. Can J For Res 39:481–487. https://doi.org/10.1139/X08-174

Bradford JB, Schlaepfer DR, Lauenroth WK, Palmquist KA (2020) Robust ecological drought projections for drylands in the 21st century. Glob Chang Biol 26:3906–3919. https://doi.org/10.1111/gcb.15075

Bradshaw CJA, Warkentin IG, Sodhi NS (2009) Urgent preservation of boreal carbon stocks and biodiversity. Trends Ecol Evol 24:541–548. https://doi.org/10.1016/j.tree.2009.03.019

Briske DD, Derner JD, Brown JR et al (2008) Rotational grazing on rangelands: reconciliation of perception and experimental evidence. Rangel Ecol Manag 61:3–17. https://doi.org/10.2111/06-159R.1

Brookshire ENJ, Stoy PC, Currey B, Finney B (2020) The greening of the Northern Great Plains and its biogeochemical precursors. Glob Chang Biol 26:5404–5413. https://doi.org/10.1111/gcb.15115

Brown S, Miltner E, Cogger C (2012) Carbon sequestration potential in urban soils. In: Lal R, Augustin B (eds) Carbon sequestration in urban ecosystems. Springer, Dordrecht, pp 173–196

Bruhwiler L, Parmentier FJW, Crill P, Leonard M, Palmer PI (2021) The Arctic carbon cycle and its response to changing climate. Curr Clim Change Rep 7:14–34. https://doi.org/10.1007/s40641-020-00169-5

Buma B, Weiss S, Hayes K, Lucash M (2020) Wildland fire reburning trends across the US West suggest only short-term negative feedback and differing climatic effects. Environ Res Lett 15:034026. https://doi.org/10.1088/1748-9326/ab6c70

Burchfield EK, Nelson KS (2021) Agricultural yield geographies in the United States. Environ Res Lett 16:054051. https://doi.org/10.1088/1748-9326/abe88d

Burke M, Driscoll A, Heft-Neal S et al (2021) The changing risk and burden of wildfire in the United States. Proc Natl Acad Sci USA 118:e2011048118. https://doi.org/10.1073/pnas.2011048118

Byrnes RC, Eastburn DJ, Tate KW, Roche LM (2018) A global meta-analysis of grazing impacts on soil health indicators. J Environ Qual 47:758–765. https://doi.org/10.2134/jeq2017.08.0313

Cai W, Prentice IC (2020) Recent trends in gross primary production and their drivers: analysis and modelling at flux-site and global scales. Environ Res Lett 15:124050. https://doi.org/10.1088/1748-9326/abc64e

Camenzind T, Hättenschwiler S, Treseder K, Lehmann A, Rillig MC (2018) Nutrient limitation of soil microbial processes in tropical forests. Ecol Monogr 88:4–21. https://doi.org/10.1002/ecm.1279

Campbell EE, Paustian K (2015) Current developments in soil organic matter modeling and the expansion of model applications: a review. Environ Res Lett 10:123004. https://doi.org/10.1088/1748-9326/10/12/123004

Carlson KM, Gerber JS, Mueller ND et al (2017) Greenhouse gas emissions intensity of global croplands. Nat Clim Chang 7:62–68. https://doi.org/10.1038/NCLIMATE3158

Chambers JQ, Fisher JI, Zeng H et al (2007) Hurricane Katrina's carbon footprint on U.S. Gulf Coast forests. Science 318:1107. https://doi.org/10.1126/science.1148913

Chang J, Ciais P, Gasser T et al (2021) Climate warming from managed grasslands cancels the cooling effect of carbon sinks in sparsely grazed and natural grasslands. Nat Commun 12:118. https://doi.org/10.1038/s41467-020-20406-7

Changnon S (2003) Characteristics of ice storms in the United States. J Appl Meteorol 42:630–639

Chaplot V (2021) Evidences of plants' impact on land degradation and climate change: an urgent call for new multidisciplinary research. Geoderma 392:114984. https://doi.org/10.1016/j.geoderma.2021.114984

Chapman BR, Bolen EG (2015) Ecology of North America, 2nd edn. Wiley, West Sussex, p 334

Chapman M, Walker WS, Cook-Patton SC et al (2020) Large climate mitigation potential from adding trees to agricultural lands. Glob Chang Biol 26:4357–4365. https://doi.org/10.1111/gcb.15121

Chen X, Chen HYH, Chen C et al (2020a) Effects of plant diversity on soil carbon in diverse ecosystems: a global meta-analysis. Biol Rev 95:167–183. https://doi.org/10.1111/brv.12554

Chen J, Elsgaard L, van Groenigen KJ et al (2020b) Soil carbon loss with warming: new evidence from carbon-degrading enzymes. Glob Chang Biol 26:1944–1952. https://doi.org/10.1111/gcb.14986

Chen J, Engbersen N, Stefan L et al (2021) Diversity increases yield but reduces harvest index in crop mixtures. Nat Plants. https://doi.org/10.1038/s41477-021-00948-4

Chimner RA, Cooper DJ, Wurster FC, Rochefort L (2017) An overview of peatland restoration in North America: where are we after 25 years? Restor Ecol 25:283–292. https://doi.org/10.1111/rec.12434

Ciampitti IA, Camberato JJ, Murrell ST, Vyn TJ (2013) Maize nutrient accumulation and partitioning in response to plant density and nitrogen rate: I. Macronutrients. Agron J 105:783–795. https://doi.org/10.2134/agronj2012.0467

Clarke N, Gundersen P, Jönsson-Belyazid U et al (2015) Influence of different tree-harvesting intensities on forest soil carbon stocks in boreal and northern temperate forest ecosystems. For Ecol Manag 351:9–19. https://doi.org/10.1016/j.foreco.2015.04.034

Conant RT, Paustian K, Elliott ET (2001) Grassland management and conversion into grassland: effects on soil carbon. Ecol Appl 11:343–355. https://doi.org/10.1890/1051-0761(2001)011[0343:GMACIG]2.0.CO;2

Conant RT, Ryan MG, Ågren GI et al (2011) Temperature and soil organic matter decomposition rates – synthesis of current knowledge and a way forward. Glob Chang Biol 17:3392–3404. https://doi.org/10.1111/j.1365-2486.2011.02496.x

Conant RT, Cerri CEP, Osborne BB, Paustian K (2017) Grassland management impacts on soil carbon stocks: a new synthesis. Ecol Appl 27:662–668. https://doi.org/10.1002/eap.1473

Cook-Patton SC, Gopalakrishna T, Daigneault A et al (2020a) Lower cost and more feasible options to restore forest cover in the contiguous United States for climate mitigation. One Earth 3:739–752. https://doi.org/10.1016/j.oneear.2020.11.013

Cook-Patton SC, Leavitt SM, Gibbs D et al (2020b) Mapping carbon accumulation potential from global natural forest regrowth. Nature 585:545–550. https://doi.org/10.1038/s41586-020-2686-x

Coop JD, Parks SA, Stevens-Rumann CS et al (2020) Wildfire-driven forest conversion in western North American landscapes. BioScience 70:659–673. https://doi.org/10.1093/biosci/biaa061

Corlett RT, Primack PB (2011) Tropical rainforests: an ecological and biogeographical comparison. Wiley, Oxford

Craig ME, Turner BL, Liang C et al (2018) Tree mycorrhizal type predicts within-site variability in the storage and distribution of soil organic matter. Glob Chang Biol 24:3317–3330. https://doi.org/10.1111/gcb.14132

Craig ME, Mayes MA, Sulman BN, Walker AP (2021) Biological mechanisms may contribute to soil carbon saturation patterns. Glob Chang Biol 27:2633–2644. https://doi.org/10.1111/gcb.15584

Crème A, Rumpel C, Malone SL et al (2020) Monitoring grassland management effects on soil organic carbon – a matter of scale. Agronomy 10:2016. https://doi.org/10.3390/agronomy10122016

Crews TE, Rumsey BE (2017) What agriculture can learn from native ecosystems in building soil organic matter: a review. Sustainability 9:578. https://doi.org/10.3390/su9040578

Crews TE, Carton W, Olsson L (2018) Is the future of agriculture perennial? Imperatives and opportunities to reinvent agriculture by shifting from annual monocultures to perennial polycultures. Glob Sustain 1:1–18. https://doi.org/10.1017/sus.2018.11

Crowley KF, McNeil BE, Lovett GM et al (2012) Do nutrient limitation patterns shift from nitrogen toward phosphorus with increasing nitrogen deposition across the Northeastern United States? Ecosystems 15:940–957. https://doi.org/10.1007/s10021-012-9550-2

Cui Y, Schubert BA, Jahren AH (2020) A 23 m.y. record of low atmospheric CO_2. Geology 48. https://doi.org/10.1130/G47681.1

Cusack DF, Marín-Spiotta E (2019) Wet tropical soils and global change. In: Busse M, Giardina CP, Morris DM, Page-Dumroese DS (eds) Global change and forest soils, Developments in soil science 36. Elsevier, Amsterdam, pp 131–169. https://doi.org/10.1016/B978-0-444-63998-1.00008-2

Dangal SRS, Tian H, Pan S, Zhang L, Xu R (2020) Environ Res Lett 15:104006. https://doi.org/10.1088/1748-9326/abaa79

Davidson EA, Janssens IA (2006) Temperature sensitivity of soil carbon decomposition and feedbacks to climate change. Nature 440:165–173. https://doi.org/10.1038/nature04514

Dean C, Kirkpatrick JB, Friedland AJ (2017) Conventional intensive logging promotes loss of organic carbon from the mineral soil. Glob Chang Biol 23:1–11. https://doi.org/10.1111/gcb.13387

Deines JM, Kendall AD, Crowley MA et al (2019) Mapping three decades of annual irrigation across the US high plains aquifer using Landsat and Google Earth Engine. Remote Sens Environ 233:111400. https://doi.org/10.1016/j.rse.2019.111400

DeLuca TH, Boisvenue C (2012) Boreal forest soil carbon: distribution, function and modelling. Forestry 85:161–184. https://doi.org/10.1093/forestry/cps003

DeLuca TH, Zabinski CA (2011) Prairie ecosystems and the carbon problem. Front Ecol Environ 9:407–413. https://doi.org/10.1890/100063

Denef K, Stewart CE, Brenner J, Paustian K (2008) Does long-term center-pivot irrigation increase soil carbon stocks in semi-arid agroecosystems? Geoderma 145:121–129. https://doi.org/10.1016/j.geoderma.2008.03.002

Derner JD, Jin VL (2012) Soil carbon dynamics and rangeland management. In: Liebig MA, Franzluebbers AJ, Follett RF (eds) Managing agricultural greenhouse gases. Elsevier, Amsterdam, pp 79–92. https://doi.org/10.1016/B978-0-12-386897-8.00006-1

Dieter CA, Maupin MA, Caldwell RR et al (2018) Estimated use of water in the United States in 2015: U.S. Geological Survey Circular 1441, 65 p. https://doi.org/10.3133/cir1441

Diochon A, Kellman L, Beltrami H (2009) Looking deeper: an investigation of soil carbon losses following harvesting from a managed northeastern red spruce (*Picea rubens* Sarg.) forest chronosequence. For Ecol Manag 257:413–420. https://doi.org/10.1016/j.foreco.2008.09.015

Dlamini P, Chivenge P, Chaplot V (2016) Overgrazing decreases soil organic carbon stocks the most under dry climates and low soil pH: a meta-analysis shows. Agric Ecosyst Environ 221:258–269. https://doi.org/10.1016/j.agee.2016.01.026

Dokoohaki H, Miguez FE, Laird D, Dumortier J (2019) Where should we apply biochar? Environ Res Lett 14:044005. https://doi.org/10.1088/1748-9326/aafcf0

Domke G, Williams CA, Birdsey R et al (2018) Chapter 9: forests. In: Cavallaro N, Shrestha G, Birdsey R et al (eds) Second state of the carbon cycle report (SOCCR2): a sustained assessment report. U.S. Global change research program, Washington, DC, pp 365–398. https://doi.org/10.7930/SOCCR2.2018.Ch9

Domke GM, Oswalt SN, Walters BF, Morin RS (2020) Tree planting has the potential to increase carbon sequestration capacity of forests in the United States. Proc Natl Acad Sci USA 117:24649–24651. https://doi.org/10.1073/pnas.2010840117

Douglas TA, Jones MC, Hiemstra CA, Arnold JR (2014) Sources and sinks of carbon in boreal ecosystems of interior Alaska: a review. Elementa 2:00032. https://doi.org/10.12952/journal.elementa.000032

Dove NC, Safford HD, Bohlman GN, Estes BL, Hart SC (2020) High-severity wildfire leads to multi-decadal impacts on soil biogeochemistry in mixed-conifer forests. Ecol Appl 30:e02072. https://doi.org/10.1002/eap.2072

Drexler JZ, Krauss KW, Sasser MC et al (2013) A long-term comparison of carbon sequestration rates in impounded and naturally tidal freshwater marshes along the lower Waccamaw River, South Carolina. Wetlands 33:965–974. https://doi.org/10.1007/s13157-013-0456-3

Du Y, Ke X, Li J et al (2021) Nitrogen deposition increases global grassland N_2O emission rates steeply: a meta-analysis. Catena 199:105105. https://doi.org/10.1016/j.catena.2020.105105

Duarte-Guardia S, Peri P, Amelung W et al (2020) Biophysical and socioeconomic factors influencing soil carbon stocks: a global assessment. Mitig Adapt Strateg Glob Chang. https://doi.org/10.1007/s11027-020-09926-1

Duffy KA, Schwalm CR, Arcus VL et al (2021) How close are we to the temperature tipping point of the terrestrial biosphere? Sci Adv 7:eaay1052. https://doi.org/10.1126/sciadv.aay1052

Dumortier J, Dokoohaki H, Elobeid A et al (2020) Global land-use and carbon emission implications from biochar application to cropland in the United States. J Clean Prod 258:120684. https://doi.org/10.1016/j.jclepro.2020.120684

Edmondson JL, Davies ZG, McHugh N, Gaston KJ, Leake JR (2012) Organic carbon hidden in urban ecosystems. Sci Rep 2:963. https://doi.org/10.1038/srep00963

Eglinton TI, Galy VV, Hemingway JD et al (2021) Climate control on terrestrial biospheric carbon turnover. Proc Natl Acad Sci USA 118:e2011585118. https://doi.org/10.1073/pnas.2011585118

Ellis EC, Ramankutty N (2008) Putting people in the map: anthropogenic biomes of the world. Front Ecol Environ 6:439–447. https://doi.org/10.1890/070062

Emde D, Hannam KD, Most I, Nelson LM, Jones MD (2021) Soil organic carbon in irrigated agricultural systems: a meta-analysis. Glob Chang Biol 27:3898–3910. https://doi.org/10.1111/gcb.15680

Entry JA, Sojka RE, Shewmaker GE (2004) Irrigation increases inorganic carbon in agricultural soils. Environ Manag 33:309–317. https://doi.org/10.1007/s00267-003-9140-3

Evans DL, Quinton JN, Davies JAC, Zhao J, Govers G (2020) Soil lifespans and how they can be extended by land use and management change. Environ Res Lett 15:0940b2. https://doi.org/10.1088/1748-9326/aba2fd

Evans CD, Peacock M, Baird AJ et al (2021) Overriding water table control on managed peatland greenhouse gas emissions. Nature 593:548–552. https://doi.org/10.1038/s41586-021-03523-1

Fargione J, Haase DL, Burney OT et al (2021) Challenges to the reforestation pipeline in the United States. Front For Glob Change 4:629198. https://doi.org/10.3389/ffgc.2021.629198

Fay PA, Prober SM, Ws H et al (2015) Grassland productivity limited by multiple nutrients. Nat Plants 1:15080. https://doi.org/10.1038/NPLANTS.2015.80

Fay PA, Hui D, Jackson RB et al (2021) Multiple constraints cause positive and negative feedbacks limiting grassland soil CO_2 efflux under CO_2 enrichment. Proc Natl Acad Sci USA 118:e2008284117. https://doi.org/10.1073/pnas.2008284117

Feliciano D, Ledo A, Hillier J, Nayak DR (2018) Which agroforestry options give the greatest soil and above ground carbon benefits in different world regions? Agric Ecosyst Environ 254:117–129. https://doi.org/10.1016/j.agee.2017.11.032

Feng Q, An C, Chen Z, Wang Z (2020) Can deep tillage enhance carbon sequestration in soils? A meta-analysis towards GHG mitigation and sustainable agricultural management. Renew Sust Energ Rev 133:110293. https://doi.org/10.1016/j.rser.2020.110293

Fisher JB, Sikka M, Oechel WC et al (2014) Carbon cycle uncertainty in the Alaskan Arctic. Biogeosciences 11:4271–4288. https://doi.org/10.5194/bg-11-4271-2014

Fisk JP, Hurtt GC, Chambers JQ et al (2013) The impacts of tropical cyclones on the net carbon balance of eastern US forests (1851–2000). Environ Res Lett 8:045017. https://doi.org/10.1088/1748-9326/8/4/045017
Fitts LA, Russell MB, Domke GM, Knight JK (2021) Modeling land use change and forest carbon stock changes in temperate forests in the United States. Carbon Balance Manage 16:20. https://doi.org/10.1186/s13021-021-00183-6
Flanagan NE, Wang H, Winton S, Richardson CJ (2020) Low-severity fire as a mechanism of organic matter protection in global peatlands: thermal alteration slows decomposition. Glob Chang Biol 26:3930–3946. https://doi.org/10.1111/gcb.15102
Follett RF, Kimble JM, Lal R (eds) (2001) The potential of U.S. grazing lands to sequester carbon and mitigate the greenhouse effect. CRC/Lewis Publishers, Boca Raton
Fox TR, Allen HL, Albaugh TJ, Rubilar R, Carlson CA (2007) Tree nutrition and forest fertilization of pine plantations in the southern United States. South J Appl For 31:5–11. https://doi.org/10.1093/sjaf/31.1.5
Franzluebbers AJ, Owens LB, Sigua GC, Cambardella CA, Haney RL (2012) Soil organic carbon under pasture management. In: Liebig MA, Franzluebbers AJ, Follett RF (eds) Managing agricultural greenhouse gases. Elsevier, Amsterdam, pp 93–110. https://doi.org/10.1016/B978-0-12-386897-8.00007-3
Franzluebbers AJ, Starks PJ, Steiner JL (2019) Conservation of soil organic carbon and nitrogen fractions in a tallgrass prairie in Oklahoma. Agronomy 9:204. https://doi.org/10.3390/agronomy9040204
Fraterrigo JM, Ream K, Knoepp JD (2018) Tree mortality from insect infestation enhances carbon stabilization in southern Appalachian forest soils. J Geophys Res Biogeosci 123:2121–2134. https://doi.org/10.1029/2018JG004431
Fusco EJ, Finn JT, Balch JK, Nagy RC, Bradley BA (2019) Invasive grasses increase fire occurrence and frequency across US ecoregions. Proc Natl Acad Sci USA 116:23594–23599. https://doi.org/10.1073/pnas.1908253116
Gamble JD, Feyereisen GW, Griffis TJ, Wente CD, Baker JM (2021) Long-term ecosystem carbon losses from silage maize-based forage cropping systems. Agric For Meteorol 306:108438. https://doi.org/10.1016/j.agrformet.2021.108438
Gao J, Carmel Y (2020) A global meta-analysis of grazing effects on plant richness. Agric Ecosyst Environ 302:107072. https://doi.org/10.1016/j.agee.2020.107072
Gao P, Terando AJ, Kupfer JA et al (2021) Robust projections of future fire probability for the conterminous United States. Sci Total Environ 789:147872. https://doi.org/10.1016/j.scitotenv.2021.147872
García-Palacios P, Crowther TW, Dacal M et al (2021) Evidence for large microbial-mediated losses of soil carbon under anthropogenic warming. Nat Rev Earth Environ 2:507–517. https://doi.org/10.1038/s43017-021-00178-4
Gass TM, Binkley D (2011) Soil nutrient losses in an altered ecosystem are associated with native ungulate grazing. J Appl Ecol 48:952–960. https://doi.org/10.1111/j.1365-2664.2011.01996.x
Gauthier S, Bernier P, Kuuluvainen T, Shvidenko AZ, Schepaschenko DG (2015) Boreal forest health and global change. Science 349:819–822. https://doi.org/10.1126/science.aaa9092
Gewin V (2020) Bringing back the bogs. Nature 578:204–208
Gherardi LA, Sala OE (2020) Global patterns and climatic controls of belowground net carbon fixation. Proc Natl Acad Sci USA 117:20038–20043. https://doi.org/10.1073/pnas.2006715117
Ghimire B, Williams CA, Collatz GJ et al (2015) Large carbon release legacy from bark beetle outbreaks across western United States. Glob Chang Biol 21:3087–3101. https://doi.org/10.1111/gcb.12933
Gilliam FS (2016) Forest ecosystems of temperate climatic regions: from ancient use to climate change. New Phytol 212:871–887. https://doi.org/10.1111/nph.14255
Gilmullina A, Rumpel C, Blagodatskaya E, Chabbi A (2020) Management of grasslands by mowing versus grazing – impacts on soil organic matter quality and microbial functioning. Appl Soil Ecol 156:103701. https://doi.org/10.1016/j.apsoil.2020.103701

Girardin CAJ, Jenkins L, Seddon N et al (2021) Nature-based solutions can help cool the planet – if we act now. Nature 593:191–194. https://doi.org/10.1038/d41586-021-01241-2

Glover JD, Culman SW, DuPont ST et al (2010a) Harvested perennial grasslands provide ecological benchmarks for agricultural sustainability. Agric Ecosyst Environ 137:3–12. https://doi.org/10.1016/j.agee.2009.11.001

Glover JD, Reganold JP, Bell LW et al (2010b) Increased food and ecosystem security via perennial grains. Science 328:1638–1639. https://doi.org/10.1126/science.1188761

Godde CM, de Boer IJM, zu Ermgassen E et al (2020) Soil carbon sequestration in grazing systems: managing expectations. Clim Chang 161:385–391. https://doi.org/10.1007/s10584-020-02673-x

Goldstein A, Turner WR, Spawn SA et al (2020) Protecting irrecoverable carbon in Earth's ecosystems. Nat Clim Chang 10:287–295. https://doi.org/10.1038/s41558-020-0738-8

Gomez-Casanovas N, Blanc-Betes E, Moore CE et al (2021) A review of transformative strategies for climate mitigation by grasslands. Sci Total Environ 799:149466. https://doi.org/10.1016/j.scitotenv.2021.149466

Graham MW, Thomas RQ, Lombardozzi DL, O'Rourke ME (2021) Modest capacity of no-till farming to offset emissions over 21st century. Environ Res Lett 16:054055. https://doi.org/10.1088/1748-9326/abe6c6

Green JK, Seneviratne SI, Berg AM et al (2019) Large influence of soil moisture on long-term terrestrial carbon uptake. Nature 565:476–479. https://doi.org/10.1038/s41586-018-0848-x

Gross CD, Harrison RB (2019) The case for digging deeper: soil organic carbon storage, dynamics, and controls in our changing world. Soil Syst 3:28. https://doi.org/10.3390/soilsystems3020028

Guenet B, Gabrielle B, Chenu C et al (2021) Can N_2O emissions offset the benefits from soil organic carbon storage? Glob Chang Biol 27:237–256. https://doi.org/10.1111/gcb.15342

Günther A, Barthelmes A, Huth V et al (2020) Prompt rewetting of drained peatlands reduces climate warming despite methane emissions. Nat Commun 11:1644. https://doi.org/10.1038/s41467-020-15499-z

Guo X, Gao Q, Yuan M et al (2020) Gene-informed decomposition model predicts lower soil carbon loss due to persistent microbial adaptation to warming. Nat Commun 11:4897. https://doi.org/10.1038/s41467-020-18706-z

Gurney KR, Romero-Lankao P, Pincetl S et al (2018) Chapter 4: understanding urban carbon fluxes. In: Cavallaro N, Shrestha G, Birdsey R et al (eds) Second state of the carbon cycle report (SOCCR2): a sustained assessment report. U.S. global change research program, Washington, DC, pp 189–228. https://doi.org/10.7930/SOCCR2.2018.Ch4

Hall J, Muscarella R, Quebbeman A et al (2020) Hurricane-induced rainfall is a stronger predictor of tropical forest damage in Puerto Rico than maximum wind speeds. Sci Rep 10:4318. https://doi.org/10.1038/s41598-020-61164-2

Halofsky JE, Peterson DL, Harvey BJ (2020) Changing wildfire, changing forests: the effects of climate change on fire regimes and vegetation in the Pacific Northwest, USA. Fire Ecol 16:4. https://doi.org/10.1186/s42408-019-0062-8

Han L, Sun K, Yang Y et al (2020) Biochar's stability and effect on the content, composition and turnover of soil organic carbon. Geoderma 364:114184. https://doi.org/10.1016/j.geoderma.2020.114184

Hao G, Zhuang Q, Pan J et al (2014) Soil thermal dynamics of terrestrial ecosystems of the conterminous United States from 1948 to 2008: an analysis with a process-based soil physical model and AmeriFlux data. Clim Chang 126:135–150. https://doi.org/10.1007/s10584-014-1196-y

Harden J, Trumbore S, Stocks B et al (2000) The role of fire in the boreal carbon budget. Glob Chang Biol 6:174–184. https://doi.org/10.1046/j.1365-2486.2000.06019.x

Harris NL, Gibbs DA, Baccini A et al (2021) Global maps of twenty-first century forest carbon fluxes. Nat Clim Chang 11:234–240. https://doi.org/10.1038/s41558-020-00976-6

Harvey S, Zhang M, Fochesatto GJ (2021) Projections of spring wheat growth in Alaska: opportunity and adaptations in a changing climate. Clim Serv 22:100235. https://doi.org/10.1016/j.cliser.2021.100235

Hatfield JL, Antle J, Garrett KA et al (2020) Indicators of climate change in agricultural systems. Clim Chang 163:1719–1732. https://doi.org/10.1007/s10584-018-2222-2

Haverd V, Smith B, Canadell JP et al (2020) Higher than expected CO_2 fertilization inferred from leaf to global observations. Glob Chang Biol 26:2390–2402. https://doi.org/10.1111/gcb.14950

He Y, Trumbore SE, Torn MS et al (2016) Radiocarbon constraints imply reduced carbon uptake by soils during the 21st century. Science 353:1419–1424. https://doi.org/10.1126/science.aad4273

He Y, Yao Y, Ji Y et al (2020) Biochar amendment boosts photosynthesis and biomass in C3 but not C4 plants: a global synthesis. GCB Bioenergy 12:605–617. https://doi.org/10.1111/gcbb.12720

He J, Chen D, Jenkins L, Loboda TV (2021) Impacts of wildfire and landscape factors on organic soil properties in Arctic tussock tundra. Environ Res Lett 16:085004. https://doi.org/10.1088/1748-9326/ac1192

Helbig M, Waddington JM, Alekseychik P et al (2020) The biophysical climate mitigation potential of boreal peatlands during the growing season. Environ Res Lett 15:104004. https://doi.org/10.1088/1748-9326/abab34

Hemes KS, Chamberlain SD, Eichelmann E, Knox SH, Baldocchi DD (2018) A biogeochemical compromise: the high methane cost of sequestering carbon in restored wetlands. Geophys Res Lett 45:6081–6091. https://doi.org/10.1029/2018GL077747

Herrick JE, Lessard VC, Spaeth KE et al (2010) National ecosystem assessments supported by scientific and local knowledge. Front Ecol Environ 8:403–408. https://doi.org/10.1890/100017

Hicke JA, Allen CD, Desai AR et al (2012) Effects of biotic disturbances on forest carbon cycling in the United States and Canada. Glob Chang Biol 18:7–34. https://doi.org/10.1111/j.1365-2486.2011.02543.x

Higuera PE, Shuman BN, Wolf KD (2021) Rocky Mountain subalpine forests now burning more than any time in recent millennia. Proc Natl Acad Sci USA 118:e2103135118. https://doi.org/10.1073/pnas.2103135118

Hisano M, Ryo M, Chen X, Chen HYH (2021) Rapid functional shifts across high latitude forests over the last 65 years. Glob Chang Biol 27:3846–3858. https://doi.org/10.1111/gcb.15710

Homer C, Dewitz J, Jin S et al (2020) Conterminous United States land cover change patterns 2001–2016 from the 2016 National Land Cover Database. ISPRS J Photogramm 162:184–199. https://doi.org/10.1016/j.isprsjprs.2020.02.019

Hoover CM (2003) Soil carbon sequestration and forest management: challenges and opportunities. In: Kimble JS, Lal R, Birdsey R, Heath LS (eds) The potential of U.S. forest soils to sequester carbon and mitigate the greenhouse effect. CRC Press LLC, Boca Raton, pp 211–256

Hoover CM, Smith JE (2021) Current aboveground live tree carbon stocks and annual net change in forests of conterminous United States. Carbon Balance Manage 16:17. https://doi.org/10.1186/s13021-021-00179-2

Horton P, Long SP, Smith P, Banwart SA, Beerling DJ (2021) Technologies to deliver food and climate security through agriculture. Nat Plants 7:250–255. https://doi.org/10.1038/s41477-021-00877-2

Hou G, Delang CO, Lu X, Gao L (2020) Grouping tree species to estimate afforestation-driven soil organic carbon sequestration. Plant Soil 455:507–518. https://doi.org/10.1007/s11104-020-04685-z

Hoy EE, Turetsky MR, Kasischke ES (2016) More frequent burning increases vulnerability of Alaskan boreal black spruce forests. Environ Res Lett 11:095001. https://doi.org/10.1088/1748-9326/11/9/095001

Hristov AN, Johnson JMF, Rice CW et al (2018) Chapter 5: agriculture. In: Cavallaro N, Shrestha G, Birdsey R et al (eds) Second state of the carbon cycle report (SOCCR2): a sus-tained assessment report. U.S. global change research program, Washington, DC, pp 229–263. https://doi.org/10.7930/SOCCR2.2018.Ch5

Huang X, Terrer C, Dijkstra FA et al (2020) New soil carbon sequestration with nitrogen enrichment: a meta-analysis. Plant Soil 454:299–310. https://doi.org/10.1007/s11104-020-04617-x

Huang J, Hartemink AE, Kucharik CJ (2021) Soil-dependent responses of US crop yields to climate variability and depth to groundwater. Agric Syst 190:103085. https://doi.org/10.1016/j.agsy.2021.103085

Humpenöder F, Karstens K, Lotze-Campen H et al (2020) Peatland protection and restoration are key for climate change mitigation. Environ Res Lett 15:104093. https://doi.org/10.1088/1748-9326/abae2a

Huntzinger DN, Michalak AM, Schwalm C et al (2017) Uncertainty in the response of terrestrial carbon sink to environmental drivers undermines carbon-climate feedback predictions. Sci Rep 7:4765. https://doi.org/10.1038/s41598-017-03818-2

ICF International (2013) Greenhouse gas mitigation options and costs for agricultural land and animal production within the United States. U.S. Department of Agriculture. https://www.usda.gov/oce/climate_change/mitigation_technologies/GHGMitigation-Production_Cost.htm

Indigo (2020) Progress report: regeneration on the American farm (July 2020). https://www.indigoag.com/progress-report-access?utm_medium=social&utm_source=twitter&utm_campaign=progress_report_2020&utm_content=progress_report_1

IPCC (2021) Summary for policymakers. In: Climate change 2021: the physical science basis. Contribution of Working Group I to the Sixth Assessment Report of the Intergovernmental Panel on Climate Change [Masson-Delmotte V, Zhai P, Pirani A et al (eds)]. Cambridge University Press. In Press

Ito A, Hajima T, Lawrence DM et al (2020) Soil carbon sequestration simulated in CMIP6-LUMIP models: implications for climatic mitigation. Environ Res Lett 15:124061. https://doi.org/10.1088/1748-9326/abc912

Iversen CM, McCormack ML, Powell AS (2017) A global fine-root ecology database to address belowground challenges in plant ecology. New Phytol 215:15–26. https://doi.org/10.1111/nph.14486

James J, Page-Dumroese D, Busse M et al (2021) Effects of forest harvesting and biomass removal on soil carbon and nitrogen: two complementary meta-analyses. For Ecol Manag 485:118935. https://doi.org/10.1016/j.foreco.2021.118935

Jandl R, Lindner M, Vesterdal L et al (2007) How strongly can forest management influence soil carbon sequestration? Geoderma 137:253–268. https://doi.org/10.1016/j.geoderma.2006.09.003

Janowiak M, Connelly WJ, Dante-Wood K et al (2017) Considering forest and grassland carbon in land management. USDA Forest Service General Technical Report WO-95

Jansson C, Faiola C, Wingler A et al (2021) Crops for carbon farming. Front Plant Sci 12:636709. https://doi.org/10.3389/fpls.2021.636709

Jaramillo VJ, Murray-Tortarolo GN (2019) Tropical dry forest soils: global change and local-scale consequences for soil biogeochemical processes. In: Busse M, Giardina CP, Morris DM, Page-Dumroese DS (eds) Global change and forest soils, Developments in soil science 36. Elsevier, Amsterdam, pp 109–130. https://doi.org/10.1016/B978-0-444-63998-1.00007-0

Jo I, Potter KM, Domke GM, Fei S (2018) Dominant forest tree mycorrhizal type mediates understory plant invasions. Ecol Lett 21:217–224. https://doi.org/10.1111/ele.12884

Johnson DW, Curtis PS (2001) Effects of forest management on soil C and N storage: meta analysis. For Ecol Manag 140:227–238

Johnson MG, Kern JS (2003) Quantifying the organic carbon held in forested soils of the United States and Puerto Rico. In: Kimble JS, Lal R, Birdsey R, Heath LS (eds) The potential of U.S. forest soils to sequester carbon and mitigate the greenhouse effect. CRC Press LLC, Boca Raton, FL, pp 47–72

Joo J, Jeong S, Zheng C et al (2020) Emergence of significant soil moisture depletion in the near future. Environ Res Lett 15:124048. https://doi.org/10.1088/1748-9326/abc6d2

Joosten H, Gaudig G, Krawczynski R et al (2015) Managing soil carbon in Europe: paludicultures as a new perspective for peatlands. In: Banwart SA, Noellemeyer E, Milne E (eds) Soil carbon: science, management and policy for multiple benefits. CAB International, Wallingford, pp 297–306

Julius SH, West JM, Nover D et al (2013) Climate change and U.S. natural resources: advancing the nation's capability to adapt. Issues Ecol 18:1–17

Jurasinski G, Ahmad S, Anadon-Rosell A et al (2020) From understanding to sustainable use of peatlands: the WETSCAPES approach. Soil Syst 4:14. https://doi.org/10.3390/soilsystems4010014

Jurgensen M, Tarpey R, Pickens J, Kolka R, Palik B (2012) Long-term effect of silvicultural thinnings on soil carbon and nitrogen pools. Soil Sci Soc Am J 76:1418–1425. https://doi.org/10.2136/sssaj2011.0257

Kane DA, Bradford MA, Fuller E, Oldfield EE, Wood SA (2021) Soil organic matter protects US maize yields and lowers crop insurance payouts under drought. Environ Res Lett 16:044018. https://doi.org/10.1088/1748-9326/abe492

Kassam A, Friedrich T, Derpsch R (2019) Global spread of conservation agriculture. Int J Environ Stud 76:29–51. https://doi.org/10.1080/00207233.2018.1494927

Keller AB, Brzostek ER, Craig ME, Fisher JB, Phillips RP (2021) Root-derived inputs are major contributors to soil carbon in temperate forests, but vary by mycorrhizal type. Ecol Lett 24:626–635. https://doi.org/10.1111/ele.13651

Kielland K, Bryant JP (1998) Moose herbivory in taiga: effects on biogeochemistry and vegetation dynamics in primary succession. Oikos 82:377–383

Kimmel K, Furey GN, Hobbie SE et al (2020) Diversity-dependent soil acidification under nitrogen enrichment constrains biomass productivity. Glob Chang Biol 26:6594–6603. https://doi.org/10.1111/gcb.15329

Kimsey MJ, Laing LE, Anderson SM et al (2020) Soil mapping, monitoring, and assessment. In: Pouyat R, Page-Dumroese D, Patel-Weynand T, Geiser L (eds) Forest and rangeland soils of the United States under changing conditions. Springer, Cham, pp 169–188. https://doi.org/10.1007/978-3-030-45216-2_9

Kishchuk B, Morris D, Lorente M et al (2016) Disturbance intensity and dominant cover type influence rate of boreal soil carbon change: a Canadian multi-regional analysis. For Ecol Manag 381:48–62. https://doi.org/10.1016/j.foreco.2016.09.002

Kleber M, Bourg IC, Coward EK et al (2021) Dynamic interactions at the mineral–organic matter interface. Nat Rev Earth Environ 2:402–421. https://doi.org/10.1038/s43017-021-00162-y

Knapp A, Conard S, Blair J (1998) Determinants of soil CO_2 flux from a sub-humid grassland: effect of fire and fire history. Ecol Appl 8:760–770

Kolka R, Trettin C, Tang W et al (2018) Chapter 13: Terrestrial wetlands. In: Cavallaro N, Shrestha G, Birdsey R et al (eds) Second state of the carbon cycle report (SOCCR2): a sus-tained assessment report. U.S. global change research program, Washington, DC, pp 507–567. https://doi.org/10.7930/SOCCR2.2018.Ch13

Kranabetter JM, Sholinder A, de Montingy L (2020) Contrasting conifer species productivity in relation to soil carbon, nitrogen and phosphorus stoichiometry of British Columbia perhumid rainforests. Biogeosciences 17:1247–1260. https://doi.org/10.5194/bg-17-1247-2020

Krause A, Arneth A, Anthoni P, Rammig A (2020) Legacy effects from historical environmental changes dominate future terrestrial carbon uptake. Earth's Future 8:e2020EF001674. https://doi.org/10.1029/2020EF001674

Krist FJ, Sapio FJ, Tkacz BM (2007) Mapping risk from forest insects and diseases. USDA, Washington, DC

Kristensen JÅ, Rousk J, Metcalfe DB (2020) Below-ground responses to insect herbivory in ecosystems with woody plant canopies: a meta-analysis. J Ecol 108:917–930. https://doi.org/10.1111/1365-2745.13319

Kurz WA, Stinson G, Rampley GJ, Dymond CC, Neilson ET (2008) Risk of natural disturbances makes future contribution of Canada's forests to the global carbon cycle highly uncertain. Proc Natl Acad Sci USA 105:1551–1555. https://doi.org/10.1073/pnas.0708133105

Kurz WA, Boisvenue C, Stinson G et al (2013) Carbon in Canada's boreal forest – a synthesis. Environ Rev 21:260–292. https://doi.org/10.1139/er-2013-0041

Kuzyakov Y, Subbotina I, Chen HQ, Bogomolova I, Xu XL (2009) Black carbon decomposition and incorporation into soil microbial biomass estimated by C-14 labeling. Soil Biol Biochem 41:210–219. https://doi.org/10.1016/j.soilbio.2008.10.016

Lacroix EM, Petrenko CL, Friedland AJ (2016) Evidence for losses from strongly bound SOM pools after clear cutting in a northern hardwood forest. Soil Sci 181:202–207. https://doi.org/10.1097/ss.0000000000000147

Lai L, Kumar S (2020) A global meta-analysis of livestock grazing impacts on soil properties. PLoS ONE 15:e0236638. https://doi.org/10.1371/journal.pone.0236638

Lajtha K, Bailey VL, McFarlane K et al (2018) Chapter 12: soils. In: Cavallaro N, Shrestha G, Birdsey R et al (eds) Second state of the carbon cycle report (SOCCR2): a sustained assessment report. U.S. global change research program, Washington, DC, pp 469–506. https://doi.org/10.7930/SOCCR2.2018.Ch12

Lal R, Reicosky DC, Hanson JD (2007) Evolution of the plow over 10,000 years and the rationale for no-till farming. Soil Tillage Res 93:1–12. https://doi.org/10.1016/j.still.2006.11.004

Lange M, Eisenhauer N, Sierra CA et al (2015) Plant diversity increases soil microbial activity and soil carbon storage. Nat Commun 6:6707. https://doi.org/10.1038/ncomms7707

Lasslop G, Hantson S, Harrison SP et al (2020) Global ecosystems and fire: multi-model assessment of fire-induced tree-cover and carbon storage reduction. Glob Chang Biol 26:5027–5041. https://doi.org/10.1111/gcb.15160

Lauriks F, Salomón RL, Steppe K (2021) Temporal variability in tree responses to elevated atmospheric CO_2. Plant Cell Environ 44:1292–1310. https://doi.org/10.1111/pce.13986

Law BE, Hudiburg TW, Berner LT et al (2018) Land use strategies to mitigate climate change in carbon dense temperate forests. Proc Natl Acad Sci USA 115:3663–3668. www.pnas.org/cgi/doi/10.1073/pnas.1720064115

Lawler JJ, Lewis DJ, Nelson E et al (2014) Projected land-use change impacts on ecosystem services in the United States. Proc Natl Acad Sci USA 111:7492–7497. https://doi.org/10.1073/pnas.1405557111

Ledo A, Smith P, Zerihun A et al (2020) Changes in soil organic carbon under perennial crops. Glob Chang Biol 26:4158–4168. https://doi.org/10.1111/gcb.15120

Lee S, Chu ML, Guzman JA, Botero-Acosta A (2021) A comprehensive modeling framework to evaluate soil erosion by water and tillage. J Environ Manag 279:11631. https://doi.org/10.1016/j.jenvman.2020.111631

Lehmann J, Joseph S (eds) (2015) Biochar for environmental management: science, technology and implementation. Routledge, New York

Lehmann J, Hansel CM, Kaiser C et al (2020) Persistence of soil organic carbon caused by functional complexity. Nat Geosci 13:529–534. https://doi.org/10.1038/s41561-020-0612-3

Lehner PH, Rosenberg NA (2019) Agriculture. In: Gerrard MB, Dernbach JC (eds) Legal pathways to deep decarbonization in the United States. Environmental Law Institute, Washington, DC, pp 772–822

Lei J, Guo X, Zeng Y et al (2021) Temporal changes in global soil respiration since 1987. Nat Commun 12:403. https://doi.org/10.1038/s41467-020-20616-z

Lembrechts JJ, Aalto J, Ashcroft MB et al (2020) SoilTemp: A global database of near-surface temperature. Glob Chang Biol 26:6616–6629. https://doi.org/10.1111/gcb.15123

Leroux SJ, Wiersma YF, Vander Wal E (2020) Herbivore impacts on carbon cycling in boreal forests. Trends Ecol Evol 35:1001–1010. https://doi.org/10.1016/j.tree.2020.07.009

Levy-Varon JH, Batterman SA, Medvigy D et al (2019) Tropical carbon sink accelerated by symbiotic dinitrogen fixation. Nat Commun 10:5637. https://doi.org/10.1038/s41467-019-13656-7

Li C, Hoffland E, Kuyper TW et al (2020) Syndromes of production in intercropping impact yield gains. Nat Plants 6:653–660. https://doi.org/10.1038/s41477-020-0680-9

Li J, Pei J, Liu J et al (2021) Spatiotemporal variability of fire effects on soil carbon and nitrogen: a global meta-analysis. Glob Chang Biol 27:4196–4206. https://doi.org/10.1111/gcb.15742

Liang W, Zhang W, Jin Z et al (2020) Estimation of global grassland net ecosystem carbon exchange using a model tree ensemble approach. J Geophys Res Biogeosci 125:e2019JG005034. https://doi.org/10.1029/2019JG005034

Liechty HO, Jurgensen MF, Mroz GD, Gale M (1997) Pit and mound topography and its influence on storage of carbon, nitrogen, and organic matter within an old-growth forest. Can J For Res 27:1992–1997

Liu M, Yang L (2021) Human-caused fires release more carbon than lightning-caused fires in the conterminous United States. Environ Res Lett 16:014013. https://doi.org/10.1088/1748-9326/abcbbc

Liu S, Zhang Y, Zong Y et al (2016) Response of soil carbon dioxide fluxes, soil organic carbon and microbial biomass carbon to biochar amendment: a meta-analysis. GCB Bioenergy 8:392–406. https://doi.org/10.1111/gcbb.12265

Liu J, Sleeter BM, Zhu Z et al (2020a) Critical land change information enhances the understanding of carbon balance in the United States. Glob Chang Biol 26:3920–3929. https://doi.org/10.1111/gcb.15079

Liu J, Wennberg PO, Parazoo NC, Yin Y, Frankenberg C (2020b) Observational constraints on the response of high-latitude northern forests to warming. AGU Adv 2:e2020AV000228. https://doi.org/10.1029/2020AV000228

Liu S, Zheng Y, Ma R et al (2020c) Increased soil release of greenhouse gases shrinks terrestrial carbon uptake enhancement under warming. Glob Chang Biol 26:4601–4613. https://doi.org/10.1111/gcb.15156

Lorenz K (2017) Managing urban soils for food production. In: Lal R, Stewart BA (eds) Advances in soil science: urban soils. CRC Press, Boca Raton, pp 298–311

Lorenz K, Lal R (2009) Biogeochemical C and N cycles in urban soils. Environ Int 35:1–8. https://doi.org/10.1016/j.envint.2008.05.006

Lorenz K, Lal R (2010) Carbon sequestration in forest ecosystems. Springer, Dordrecht

Lorenz K, Lal R (2014a) Soil organic carbon sequestration in agroforestry systems. A review. Agron Sustain Dev 34:443–454. https://doi.org/10.1007/s13593-014-0212-y

Lorenz K, Lal R (2014b) Biochar application to soil for climate change mitigation by soil organic carbon sequestration. J Plant Nutr Soil Sci 177:651–670. https://doi.org/10.1002/jpln.201400058

Lorenz K, Lal R (2015) Managing soil carbon stocks to enhance the resilience of urban ecosystems. Carbon Manage 6:35–50. https://doi.org/10.1080/17583004.2015.1071182

Lorenz K, Lal R (2017) Impacts of land take and soil sealing on soil carbon. In: Gardi C (ed) Urban expansion, land cover and soil ecosystem services. Taylor & Francis, Routledge, Abingdon, pp 181–192

Lorenz K, Shaw RK (2017) Carbon storage in urban soils. In: Levine MJ, Kim KHJ, Morel JL, Burghardt W, Charzynski P, Shaw RK (eds) Soils within cities. Catena-Schweizerbart, Stuttgart, pp 203–205

Lu J, Carbone GJ, Huang X, Lackstrom K, Gao P (2020) Mapping the sensitivity of agriculture to drought and estimating the effect of irrigation in the United States, 1950–2016. Agric For Meteorol 292–293:108124. https://doi.org/10.1016/j.agrformet.2020.108124

Lu X, Vitousek PM, Mao Q et al (2021) Nitrogen deposition accelerates soil carbon sequestration in tropical forests. Proc Natl Acad Sci USA 118:e2020790118. https://doi.org/10.1073/pnas.2020790118

Lucchese M, Waddington JM, Poulin M et al (2010) Organic matter accumulation in a restored peatland: Evaluating restoration success. Ecol Eng 36:482–488. https://doi.org/10.1016/j.ecoleng.2009.11.017

Lugato E, Leip A, Jones A (2018) Mitigation potential of soil carbon management overestimat-ed by neglecting N_2O emissions. Nat Clim Chang 8:219–223. https://doi.org/10.1038/s41558-018-0087-z

Lugato E, Lavallee JM, Haddix ML, Panagos P, Cotrufo MF (2021) Different climate sensitivity of particulate and mineral-associated soil organic matter. Nat Geosci 14:295–300. https://doi.org/10.1038/s41561-021-00744-x

Luo Y, Ahlström A, Allison SD et al (2016) Toward more realistic projections of soil carbon dynamics by Earth system models. Glob Biogeochem Cycles 30:40–56. https://doi.org/10.1002/2015GB005239

Luo Z, Rossel RAV, Shi Z (2020) Distinct controls over the temporal dynamics of soil carbon fractions after land use change. Glob Chang Biol 26:4614–4625. https://doi.org/10.1111/gcb.15157

Lyu Z, Genet H, He Y et al (2018) The role of environmental driving factors in historical and projected carbon dynamics of wetland ecosystems in Alaska. Ecol Appl 28:1377–1395

Mack MC, Schuur EAG, Bret-Harte MS, Shaver GR, Chapin FS (2004) Ecosystem carbon storage in arctic tundra reduced by long-term nutrient fertilization. Nature 431:440–443. https://doi.org/10.1038/nature02887

Mack MC, Walker XJ, Johnstone JF et al (2021) Carbon loss from boreal forest wildfires offset by increased dominance of deciduous trees. Science 372:280–283. https://doi.org/10.1126/science.abf3903

Madani N, Parazoo NC, Kimball JS et al (2020) Recent amplified global gross primary productivity due to temperature increase is offset by reduced productivity due to water constraints. AGU Adv 2:e2020AV000180. https://doi.org/10.1029/2020AV000180

Madrigal-González J, Calatayud J, Ballesteros-Cánovas JA et al (2020) Climate reverses directionality in the richness–abundance relationship across the World's main forest biomes. Nat Commun 11:5635. https://doi.org/10.1038/s41467-020-19460-y

Marcolla B, Migliavacca M, Rödenbeck C, Cescatti A (2020) Patterns and trends of the dominant environmental controls of net biome productivity. Biogeosciences 17:2365–2379. https://doi.org/10.5194/bg-17-2365-2020

Martinuzzi S, Gavier-Pizarro GI, Lugo AE, Radeloff VC (2015) Future land-use changes and the potential for novelty in ecosystems of the United States. Ecosystems 18:1332–1342. https://doi.org/10.1007/s10021-015-9901-x

Mathew I, Shimelis H, Mutema M, Minasny B, Chaplot V (2020) Crops for increasing soil organic carbon stocks – a global meta analysis. Geoderma 367:114230. https://doi.org/10.1016/j.geoderma.2020.114230

Matthes JH, Lang AK, Jevon FV, Russell SJ (2018) Tree stress and mortality from Emerald Ash Borer does not systematically alter short-term soil carbon flux in a mixed northeastern U.S. forest. Forests 9:37. https://doi.org/10.3390/f9010037

Maurer GE, Hallmark AJ, Brown RF, Sala OE, Collins SL (2020) Sensitivity of primary production to precipitation across the United States. Ecol Lett 23:527–536. https://doi.org/10.1111/ele.13455

Mayer M, Prescott CE, Abaker WEA et al (2020) Tamm review: influence of forest management activities on soil organic carbon stocks: a knowledge synthesis. For Ecol Manag 466:118127. https://doi.org/10.1016/j.foreco.2020.118127

McCarthy MI, Ramsey B, Phillips J, Redsteer MH (2018) Chapter 7: tribal lands. In: Cavallaro N, Shrestha G, Birdsey R et al (eds) Second state of the carbon cycle report (SOCCR2): a sustained assessment report. U.S. global change research program, Washington, DC, pp 303–335. https://doi.org/10.7930/SOCCR2.2018.Ch7

McDowell NG, Allen CD, Anderson-Teixeira K et al (2020) Pervasive shifts in forest dynamics in a changing world. Science 368:eaaz9463. https://doi.org/10.1126/science.aaz9463

McGuire AD, Chapin FS, Walsh JE, Wirth C (2006) Integrated regional changes in Arctic climate feedbacks: implications for the global climate system. Annu Rev Environ Resour 31:61–91. https://doi.org/10.1146/annurev.energy.31.020105.100253

McSherry ME, Ritchie ME (2013) Effects of grazing on grassland soil carbon: a global review. Glob Chang Biol 19:1347–1357. https://doi.org/10.1111/gcb.12144

Meier EA, Thorburn PJ, Bell LW, Harrison MT, Biggs JS (2020) Greenhouse gas emissions from cropping and grazed pastures are similar: a simulation analysis in Australia. Front Sustain Food Syst 3:121. https://doi.org/10.3389/fsufs.2019.00121

Melillo JM, Frey SD, DeAngelis KM et al (2017) Long-term pattern and magnitude of soil carbon feedback to the climate system in a warming world. Science 358:101–105. https://doi.org/10.1126/science.aan2874

Melvin AM, Lichstein JW, Goodale CL (2013) Forest liming increases forest floor carbon and nitrogen stocks in a mixed hardwood forest. Ecol Appl 23:1962–1975. https://doi.org/10.1890/13-0274

Menichetti L, Kätterer T, Bolinder MA (2020) A Bayesian modeling framework for estimating equilibrium soil organic C sequestration in agroforestry systems. Agric Ecosyst Environ 303:107118. https://doi.org/10.1016/j.agee.2020.107118

Minasny B, McBratney AB, Malone BP, Wheeler I (2013) Digital mapping of soil carbon. Adv Agron 118:1–47. https://doi.org/10.1016/B978-0-12-405942-9.00001-3

Minkkinen K, Korhonen R, Savolainen I, Laine J (2002) Carbon balance and radiative forcing of Finnish peatlands 1900–2100 – the impact of forestry drainage. Glob Chang Biol 8:785–799. https://doi.org/10.1046/j.1365-2486.2002.00504.x

Mobley ML, Lajtha K, Kramer MG et al (2015) Surficial gains and subsoil losses of soil carbon and nitrogen during secondary forest development. Glob Chang Biol 21:986–996. https://doi.org/10.1111/gcb.12715

Moreno-Mateos D, Power ME, Comín FA, Yockteng R (2012) Structural and functional loss in restored wetland ecosystems. PLoS Biol 10:e1001247. https://doi.org/10.1371/journal.pbio.1001247

Mosier S, Apfelbaum S, Byck P et al (2021) Adaptive multi-paddock grazing enhances soil carbon and nitrogen stocks and stabilization through mineral association in southeastern U.S. grazing lands. J Environ Manag 288:112409. https://doi.org/10.1016/j.jenvman.2021.112409

Mwafulirwa L, Paterson E, Cairns JE et al (2021) Genotypic variation in maize (*Zea mays*) influences rates of soil organic matter mineralization and gross nitrification. New Phytol 231:2015–2028. https://doi.org/10.1111/nph.17537

Nave LE, Vance ED, Swanston CW, Curtis PS (2010) Harvest impacts on soil carbon storage in temperate forests. For Ecol Manag 259:857–866. https://doi.org/10.1016/j.foreco.2009.12.009

Nave LE, Vance ED, Swanston CW, Curtis PS (2011) Fire effects on temperate forest soil C and N storage. Ecol Appl 21:1189–1201. https://doi.org/10.1890/10-0660.1

Nave LE, Domke GM, Hofmeister KL et al (2018) Reforestation can sequester two petagrams of carbon in US topsoils in a century. Proc Natl Acad Sci USA 115:2776–2781. https://doi.org/10.1073/pnas.1719685115

Nave L, Marín-Spiotta E, Ontl T, Peters M, Swanston C (2019a) Soil carbon management. In: Busse M, Giardina CP, Morris DM, Page-Dumroese DS (eds) Global change and forest soils, Developments in soil science 36. Elsevier, Amsterdam, pp 215–257. https://doi.org/10.1016/B978-0-444-63998-1.00011-2

Nave LE, Walters BF, Hofmeister KL et al (2019b) The role of reforestation in carbon sequestration. New For 50:115–137. https://doi.org/10.1007/s11056-018-9655-3

Nelson KS, Burchfield EK (2021) Landscape complexity and US crop production. Nat Plants 2:330–338. https://doi.org/10.1038/s43016-021-00281-1

Nicoloso RS, Rice CW (2021) Intensification of no-till agricultural systems: an opportunity for carbon sequestration. Soil Sci Soc Am J. https://doi.org/10.1002/saj2.20260

Norby RJ, Zak DR (2011) Ecological lessons from free-air CO_2 enrichment (FACE) experiments. Annu Rev Ecol Evol Syst 42:181–203. https://doi.org/10.1146/annurev-ecolsys-102209-144647

Norris T, Vines PL, Hoeffel EM (2012) The American Indian and Alaska Native population: 2010 census briefs. U.S. Census Bureau

North MP, Hurteau MD (2011) High-severity wildfire effects on carbon stocks and emissions in fuels treated and untreated forest. For Ecol Manag 261:1115–1120. https://doi.org/10.1016/j.foreco.2010.12.039

Nunes MR, van Es HM, Veum KS, Amsili JP, Karlen DL (2020a) Anthropogenic and inherent effects on soil organic carbon across the U.S. Sustainability 12:5695. https://doi.org/10.3390/su12145695

Nunes MR, Karlen DL, Moorman TB, Cambardella CA (2020b) How does tillage intensity affect chemical soil health indicators? A United States meta-analysis. Agrosyst Geosci Environ 3:e20083. https://doi.org/10.1002/agg2.20083

Nunes MR, Karlen DL, Veum KS, Moorman TB, Cambardella CA (2020c) Biological soil health indicators respond to tillage intensity: a US meta-analysis. Geoderma 369:114335. https://doi.org/10.1016/j.geoderma.2020.114335

Ochoa-Hueso R, Borer ET, Seabloom EW et al (2020) Microbial processing of plant remains is co-limited by multiple nutrients in global grasslands. Glob Chang Biol 26:4572–4582. https://doi.org/10.1111/gcb.15146

Ogle SM, Conant RT, Paustian K (2004) Deriving grassland management factors for a carbon accounting method developed by the intergovernmental panel on climate change. Environ Manag 33:474–484. https://doi.org/10.1007/s00267-003-9105-6

Ogle SM, Alsaker C, Baldock J et al (2019) Climate and soil characteristics determine where no-till management can store carbon in soils and mitigate greenhouse gas emissions. Sci Rep 9:11665. https://doi.org/10.1038/s41598-019-47861-7

Olson SK, Smithwick EAH, Lucash MS et al (2021) Landscape-scale forest reorganization following insect invasion and harvest under future climate change scenarios. Ecosystems. https://doi.org/10.1007/s10021-021-00616-w

Oswalt SN, Smith WB, Miles PD et al (2019) Forest resources of the United States, 2017: a technical document supporting the Forest Service 2020 RPA Assessment. Gen Tech Rep WO-97. U.S. Department of Agriculture, Forest Service, Washington Office, Washington, DC, 223 p. https://doi.org/10.2737/WO-GTR-97

Overby ST, Hart SC, Neary DG (2003) Impacts of natural disturbances on soil carbon dynamics in forest ecosystems. In: Kimble JS, Lal R, Birdsey R, Heath LS (eds) The potential of U.S. forest soils to sequester carbon and mitigate the greenhouse effect. CRC Press LLC, Boca Raton, pp 159–172

Page KL, Dang YP, Dalal RC (2020) The ability of conservation agriculture to conserve soil organic carbon and the subsequent impact on soil physical, chemical, and biological properties and yield. Front Sustain Food Syst 4:31. https://doi.org/10.3389/fsufs.2020.00031

Page-Dumroese D, Jurgensen MF, Harvey AE (2003) Fire and fire-suppression impacts on forest-soil carbon. In: Kimble JS, Lal R, Birdsey R, Heath LS (eds) The potential of U.S. forest soils to sequester carbon and mitigate the greenhouse effect. CRC Press LLC, Boca Raton, pp 201–210

Palviainen M, Laurén A, Pumpanen J et al (2020) Decadal-scale recovery of carbon stocks after wildfires throughout the boreal forests. Glob Biogeochem Cycles 34:e2020GB006612. https://doi.org/10.1029/2020GB006612

Paustian K, Larson E, Kent J, Marx E, Swan A (2019) Soil C sequestration as a biological negative emission strategy. Front Clim 1:8. https://doi.org/10.3389/fclim.2019.00008

Pedersen P, Lauer JG (2004) Response of soybean yield components to management system and planting date. Agron J 96:1372–1381. https://doi.org/10.2134/agronj2004.1372

Peixoto L, Elsgaard L, Rasmussen J et al (2020) Decreased rhizodeposition, but increased microbial carbon stabilization with soil depth down to 3.6 m. Soil Biol Biochem 150:108008. https://doi.org/10.1016/j.soilbio.2020.108008

Pellegrini AFA, Ahlström A, Hobbie SE et al (2018) Fire frequency drives decadal changes in soil carbon and nitrogen and ecosystem productivity. Nature 553:194–198. https://doi.org/10.1038/nature24668

Pellegrini AFA, Caprio AC, Georgiou K et al (2021) Low-intensity frequent fires in coniferous forests transform soil organic matter in ways that may offset ecosystem carbon losses. Glob Chang Biol 27:3810–3823. https://doi.org/10.1111/gcb.15648

Pendall E, Bachelet D, Conant RT et al (2018) Chapter 10: grasslands. In: Cavallaro N, Shrestha G, Birdsey R et al (eds) Second state of the carbon cycle report (SOCCR2): a sustained assessment report. U.S. global change research program, Washington, DC, pp 399–427. https://doi.org/10.7930/SOCCR2.2018.Ch10

Peng Y, Schmidt IK, Zheng H et al (2020) Tree species effects on topsoil carbon stock and concentration are mediated by tree species type, mycorrhizal association, and N-fixing ability at the global scale. For Ecol Manag 478:118510. https://doi.org/10.1016/j.foreco.2020.118510

Persson T, Andersson S, Bergholm J et al (2021) Long-term impact of liming on soil C and N in a fertile spruce forest ecosystem. Ecosystems 24:968–987. https://doi.org/10.1007/s10021-020-00563-y

Petrenko CL, Friedland AJ (2015) Mineral soil carbon pool responses to forest clearing in Northeastern hardwood forests. GCB Bioenergy 7:1283–1293. https://doi.org/10.1111/gcbb.12221

Phillips CL (2020) How much will soil warm? J Geophys Res Biogeosci 125:e2020JG005668. https://doi.org/10.1029/2020JG005668

Ping CL, Michaelson GJ, Kane ES et al (2010) Carbon stores and biogeochemical properties of soils under black spruce forest, Alaska. Soil Sci Soc Am J 74:969–978. https://doi.org/10.2136/sssaj2009.0152

Poeplau C, Don A (2015) Carbon sequestration in agricultural soils via cultivation of cover crops – a meta-analysis. Agric Ecosyst Environ 200:33–41. https://doi.org/10.1016/j.agee.2014.10.024

Powers RF, Busse MD, McFarlane KJ, Zhang J, Young DH (2013) Long-term effects of silviculture on soil carbon storage: does vegetation control make a difference? Forestry 86:47–58. https://doi.org/10.1093/forestry/cps067

Prăvălie R (2018) Major perturbations in the Earth's forest ecosystems. Possible implications for global warming. Earth Sci Rev 185:544–571. https://doi.org/10.1016/j.earscirev.2018.06.010

Prăvălie R (2021) Exploring the multiple land degradation pathways across the planet. Earth Sci Rev 220:103689. https://doi.org/10.1016/j.earscirev.2021.103689

Price SP, Bradford MA, Ashton MS (2012) Characterizing organic carbon stocks and flows in forest soils. In: Ashton MS, Tyrrell ML, Spalding D, Gentry B (eds) Managing forest carbon in a changing climate. Springer, Dordrecht, pp 7–30. https://doi.org/10.1007/978-94-007-2232-3_2

Pries CEH, Castanha C, Porras RC, Torn MS (2017) The whole-soil carbon flux in response to warming. Science 355:1420–1423. https://doi.org/10.1126/science.aal1319

Ramirez JI, Jansen PA, Poorter L (2018) Effects of wild ungulates on the regeneration, structure and functioning of temperate forests: a semi-quantitative review. For Ecol Manag 424:406–419. https://doi.org/10.1016/j.foreco.2018.05.016

Rasse DP, Rumpel C, Dignac M-F (2005) Is soil carbon mostly root carbon? Mechanisms for a specific stabilization. Plant Soil 269:341–356. https://doi.org/10.1007/s11104-004-0907-y

Raymer PC, Orwig DA, Finzi AC (2013) Hemlock loss due to the hemlock woolly adelgid does not affect ecosystem C storage but alters its distribution. Ecosphere 4:63. https://doi.org/10.1890/ES12-00362.1

Reicosky DC (2015) Conservation tillage is not conservation agriculture. J Soil Water Conserv 70:103A–108A. https://doi.org/10.2489/jswc.70.5.103A

Renforth P, Edmondson J, Leake JR, Gaston KJ, Manning DAC (2011) Designing a carbon capture function into urban soils. Proc Inst Civil Eng Urban Design Plan 164:121–128

Ritchie ME (2020) Grazing management, forage production and soil carbon dynamics. Resources 9:49. https://doi.org/10.3390/resources9040049

Rogers BM, Balch JK, Goetz SJ, Lehmann CER, Turetsky M (2020) Focus on changing fire regimes: interactions with climate, ecosystems, and society. Environ Res Lett 15:030201. https://doi.org/10.1088/1748-9326/ab6d3a

Rolando JL, Dubeux JCB, de Souza TC et al (2021) Organic carbon is mostly stored in deep soil and only affected by land use in its superficial layers: a case study. Agrosyst Geosci Environ 4:e20135. https://doi.org/10.1002/agg2.20135

Ryals R, Hartman MD, Parton WJ, DeLonge MS, Silver WL (2015) Long-term climate change mitigation potential with organic matter management on grasslands. Ecol Appl 25:531–545. https://doi-org.proxy.lib.ohio-state.edu/10.1890/13-2126.1

Ryan MG, Harmon ME, Birdsey RA et al (2010) A synthesis of the science on forests and carbon for U.S. forests. Issues Ecol 13:1–16

Saban JM, Watson-Lazowski A, Chapman MA, Taylor G (2020) The methylome is altered for plants in a high CO_2 world: insights into the response of a wild plant population to multigenerational exposure to elevated atmospheric [CO_2]. Glob Chang Biol 26:6474–6492. https://doi.org/10.1111/gcb.15249

Safford HD, Vallejo VR (2019) Ecosystem management and ecological restoration in the Anthropocene: integrating global change, soils, and disturbance in boreal and Mediterranean forests. In: Busse M, Giardina CP, Morris DM, Page-Dumroese DS (eds) Global change and forest soils, Developments in soil science 36. Elsevier, Amsterdam, pp 259–306. https://doi.org/10.1016/B978-0-444-63998-1.00012-4

Sainju UM (2020) No-till Farming systems in North America. In: Dang YP, Dalal RC, Menzies NW (eds) No-till farming systems for sustainable agriculture. Springer Nature, Cham, pp 587–599. https://doi.org/10.1007/978-3-030-46409-7_32

Sanderman J, Hengl T, Fiske GJ (2017) Soil carbon debt of 12,000 years of human land use. Proc Natl Acad Sci USA 114:9575–9580. https://doi.org/10.1073/pnas.1706103114

Sanderson JS, Beutler C, Brown JR et al (2020) Cattle, conservation, and carbon in the western Great Plains. J Soil Water Conserv 75:5A–12A. https://doi.org/10.2489/jswc.75.1.5A

Santín C, Doerr SH (2016) Fire effects on soils: the human dimension. Philos Trans R Soc B 371:20150171. https://doi.org/10.1098/rstb.2015.0171

Sapkota A, Haghverdi A, Avila CCE, Ying SC (2020) Irrigation and greenhouse gas emissions: a review of field-based studies. Soil Syst 4:20. https://doi.org/10.3390/soilsystems4020020

Sarkar R, Corriher-Olson V, Long C, Somenahally A (2020) Challenges and potentials for soil organic carbon sequestration in forage and grazing systems. Rangel Ecol Manag. https://doi.org/10.1016/j.rama.2020.04.002

Sayer EJ, Lopez-Sangil L, Crawford JA et al (2019) Tropical forest soil carbon stocks do not increase despite 15 years of doubled litter inputs. Sci Rep 9:18030. https://doi.org/10.1038/s41598-019-54487-2

Scharenbroch B, Bialecki MB, Fahey RT (2017) Distribution and factors controlling soil organic carbon in the Chicago Region, Illinois, USA. Soil Sci Soc Am J 81:1436–1449. https://doi.org/10.2136/sssaj2017.03.0087

Schiedung M, Bellè SL, Sigmund G, Kalbitz K, Abiven S (2020) Vertical mobility of pyrogenic organic matter in soils: a column experiment. Biogeosciences 17:6457–6474. https://doi.org/10.5194/bg-17-6457-2020

Schlautman B, Bartel C, Diaz-Garcia L et al (2021) Perennial groundcovers: an emerging technology for soil conservation and the sustainable intensification of agriculture. Emerg Topics Life Sci. https://doi.org/10.1042/ETLS20200318

Schoeneberger MM, Bentrup G, Patel-Weynand T (eds) (2017) Agroforestry: enhancing resiliency in U.S. agricultural landscapes under changing conditions. USDA Forest Service Gen. Tech. Report WO-96

Scholten RC, Jandt R, Miller EA, Rogers BM, Veraverbeke S (2021) Overwintering fires in boreal forests. Nature 593:399–404. https://doi.org/10.1038/s41586-021-03437-y

Schulte-Uebbing L, de Vries W (2018) Global-scale impacts of nitrogen deposition on tree carbon sequestration in tropical, temperate, and boreal forests: a meta-analysis. Glob Chang Biol 24:e416–e431. https://doi.org/10.1111/gcb.13862

Schulz K, Voigt K, Beusch C et al (2016) Grazing deteriorates the soil carbon stocks of Caatinga forest ecosystems in Brazil. For Ecol Manag 367:62–70. https://doi.org/10.1016/j.foreco.2016.02.011

Schuur EAG, McGuire AD, Romanovsky V, Schädel C, Mack M (2018) Chapter 11: arctic and boreal carbon. In: Cavallaro N, Shrestha G, Birdsey R et al (eds) Second state of the carbon cycle report (SOCCR2): a sustained assessment report. U.S. global change research program, Washington, DC, pp 428–468. https://doi.org/10.7930/SOCCR2.2018.Ch11

Seidl R, Thom D, Kautz M et al (2017) Forest disturbances under climate change. Nat Clim Chang 7:395–402. https://doi.org/10.1038/NCLIMATE3303

Seifert CA, Azzari G, Lobell DB (2018) Satellite detection of cover crops and their effects on crop yield in the Midwestern United States. Environ Res Lett 13:064033. https://doi.org/10.1088/1748-9326/aac4c8

Selhorst AL, Lal R (2013) Net carbon sequestration potential and emission in home lawn turfgrasses of the United States. Environ Manag 51:198–208. https://doi.org/10.1007/s00267-012-9967-6

Sierra CA, Crow SE, Heimann M, Metzler H, Schulze ED (2021) The climate benefit of carbon sequestration. Biogeosciences 18:1029–1048. https://doi.org/10.5194/bg-18-1029-2021

Silver WL, Lugo AE, Farmer D (2003) Soil organic carbon in tropical forests of the United States of America. In: Kimble JS, Lal R, Birdsey R, Heath LS (eds) The potential of U.S. forest soils to sequester carbon and mitigate the greenhouse effect. CRC Press LLC, Boca Raton, pp 363–382

Sitters J, Wubs ERJ, Bakkers ES et al (2020) Nutrient availability controls the impact of mammalian herbivores on soil carbon and nitrogen pools in grasslands. Glob Chang Biol 26:2060–2071. https://doi.org/10.1111/gcb.15023

Skinner RH (2013) Nitrogen fertilization effects on pasture photosynthesis, respiration, and ecosystem carbon content. Agric Ecosyst Environ 172:35–41. https://doi.org/10.1016/j.agee.2013.04.005

Smethurst PJ (2010) Forest fertilization: trends in knowledge and practice compared to agriculture. Plant Soil 335:83–100. https://doi.org/10.1007/s11104-010-0316-3

Smith P (2016) Soil carbon sequestration and biochar as negative emission technologies. Glob Chang Biol 22:1315–1324. https://doi.org/10.1111/gcb.13178

Smith P, Martino D, Cai Z et al (2008) Greenhouse gas mitigation in agriculture. Philos Trans R Soc B 363:789–813. https://doi.org/10.1098/rstb.2007.2184

Smith MM, Bentrup G, Kellerman T et al (2021) Windbreaks in the United States: a systematic review of producer-reported benefits, challenges, management activities and drivers of adoption. Agric Syst 187:103032. https://doi.org/10.1016/j.agsy.2020.103032

Sollenberger LE, Kohmann MM, Dubeux JCB, Silveira ML (2019) Grassland management affects delivery of regulating and supporting ecosystem services. Crop Sci 59:441–459. https://doi.org/10.2135/cropsci2018.09.0594

Song XP, Hansen MC, Stehman SV et al (2018) Global land change from 1982 to 2016. Nature 560:639–643. https://doi.org/10.1038/s41586-018-0411-9

Soong JL, Cotrufo MF (2015) Annual burning of a tallgrass prairie inhibits C and N cycling in soil, increasing recalcitrant pyrogenic organic matter storage while reducing N availability. Glob Chang Biol 21:2321–2333. https://doi.org/10.1111/gcb.12832

Soong JL, Phillips CL, Ledna C, Koven CD, Torn MS (2020) CMIP5 models predict rapid and deep soil warming over the 21st century. J Geophys Res Biogeosci 125:e2019JG005266. https://doi.org/10.1029/2019JG005266

Soong JL, Castanha C, Pries CEH et al (2021) Five years of whole-soil warming led to loss of subsoil carbon stocks and increased CO_2 efflux. Sci Adv 7:eabd1343. https://doi.org/10.1126/sciadv.abd1343

Soussana JF, Lemaire G (2014) Coupling carbon and nitrogen cycles for environmentally sustainable intensification of grasslands and crop-livestock systems. Agric Ecosyst Environ 120:9–17. https://doi.org/10.1016/j.agee.2013.10.012

Soussana JF, Loiseau P, Vuichard N et al (2004) Carbon cycling and sequestration opportunities in temperate grasslands. Soil Use Manag 20:219–230. https://doi.org/10.1079/SUM2003234

Spawn SA, Lark TJ, Gibbs HK (2018) Carbon emissions from cropland expansion in the United States. Environ Res Lett 14:045009. https://doi.org/10.1088/1748-9326/ab0399

Sperow M (2020a) Updated potential soil carbon sequestration rates on U.S. agricultural land based on the 2019 IPCC guidelines. Soil Tillage Res 204:104719. https://doi.org/10.1016/j.still.2020.104719

Sperow M (2020b) What might it cost to increase soil organic carbon using no-till on U.S. cropland? Carbon Balance. Manage 15(26). https://doi.org/10.1186/s13021-020-00162-3

Staccone AP, Kou-Giesbrecht S, Taylor BN, Menge DNL (2021) Nitrogen-fixing trees have no net effect on forest growth in the coterminous United States. J Ecol 109:877–887. https://doi.org/10.1111/1365-2745.13513

Steele R, Hatfield JL (2018) Navigating climate-related challenges on working lands: a special issue by the USDA Climate Hubs and their partners. Clim Chang 146:1–3. https://doi.org/10.1007/s10584-017-2129-3

Steiner JL, Starks PJ, Neel JP et al (2019) Managing tallgrass prairies for productivity and ecological function: a long-term grazing experiment in the Southern Great Plains, USA. Agronomy 9:699. https://doi.org/10.3390/agronomy9110699

Stewart WM, Dibb DW, Johnston AE, Smyth TJ (2005) The contribution of commercial fertilizer nutrients to food production. Agron J 97:1–6

Stirling E, Fitzpatrick RW, Mosley LM (2020) Drought effects on wet soils in inland wetlands and peatlands. Earth Sci Rev 210:103387. https://doi.org/10.1016/j.earscirev.2020.103387

Stritar ML, Schweitzer JA, Hart SC, Bailey JK (2010) Introduced ungulate herbivore alters soil processes after fire. Biol Invasions 12:313–324. https://doi.org/10.1007/s10530-009-9624-z

Sun W, Canadell J, Yu L et al (2020) Climate drives global soil carbon sequestration and crop yield changes under conservation agriculture. Glob Chang Biol 26:3325–3335. https://doi.org/10.1111/gcb.15001

Swan A, Williams SA, Brown K et al (2015) COMET-Planner. Carbon and greenhouse gas evaluation for NRCS conservation practice planning. A companion report to www.comet-planner.com. http://comet-planner.nrel.colostate.edu/COMET-Planner_Report_Final.pdf

Sweet SK, Schuldt JP, Lehmann J, Bossio DA, Woolf D (2021) Perceptions of naturalness predict US public support for Soil Carbon Storage as a climate solution. Clim Chang 166:22. https://doi.org/10.1007/s10584-021-03121-0

Tamburini G, Bommarco R, Wanger TC et al (2020) Agricultural diversification promotes multiple ecosystem services without compromising yield. Sci Adv 6:eaba1715. https://doi.org/10.1126/sciadv.aba1715

Tanentzap AJ, Coomes DA (2012) Carbon storage in terrestrial ecosystems: do browsing and grazing herbivores matter? Biol Rev 87:72–94. https://doi.org/10.1111/j.1469-185X.2011.00185.x

Tellatin S, Myers RL (2018) Cover crop impacts on US cropland carbon sequestration. J Soil Water Conserv 73:117A–121A. https://doi.org/10.2489/jswc.73.5.117A

Terrer C, Phillips RP, Hungate BA et al (2021) A trade-off between plant and soil carbon storage under elevated CO_2. Nature 591:599–603. https://doi.org/10.1038/s41586-021-03306-8

Thaler EA, Larsen IJ, Yu Q (2021) The extent of soil loss across the US Corn Belt. Proc Natl Acad Sci USA 118:e1922375118. https://doi.org/10.1073/pnas.1922375118

Thiffault E (2019) Boreal forests and soils. In: Busse M, Giardina CP, Morris DM, Page-Dumroese DS (eds) Global change and forest soils, Developments in soil science 36. Elsevier, Amsterdam, pp 59–82. https://doi.org/10.1016/B978-0-444-63998-1.00005-7

Thiffault E, Hannam KD, Paré D et al (2011) Effects of forest biomass harvesting on soil productivity in boreal and temperate forests – a review. Environ Rev 19:278–309. https://doi.org/10.1139/A11-009

Thompson GL, Kao-Kniffin J (2017) Applying biodiversity and ecosystem function theory to turfgrass management. Crop Sci 57:S-238–S-248. https://doi.org/10.2135/cropsci2016.05.0433

Thompson JA, Kienast-Brown S, D'Avello T, Philippe J, Brungard C (2020) Soils2026 and digital soil mapping – a foundation for the future of soils information in the United States. Geoderma Reg 22:e00294. https://doi.org/10.1016/j.geodrs.2020.e00294

Tian H, Lu C, Yang J et al (2015) Global patterns and controls of soil organic carbon dynamics as simulated by multiple terrestrial biosphere models: current status and future directions. Glob Biogeochem Cycles 29:775–792. https://doi.org/10.1002/2014GB005021

Tiefenbacher A, Sandén T, Haslmayr HP et al (2021) Optimizing carbon sequestration in croplands: a synthesis. Agronomy 11:882. https://doi.org/10.3390/agronomy11050882

Tiller VE (1995) Tiller's guide to Indian country: economic profiles of American Indian reservations. BowArrow Publishing Company, Albuquerque

Tilman D (2020) Benefits of intensive agricultural intercropping. Nat Plants 6:604–605. https://doi.org/10.1038/s41477-020-0677-4

Toberman H, Laiho R, Evans CD et al (2010) Long-term drainage for forestry inhibits extracellular phenol oxidase activity in Finnish boreal mire peat. Eur J Soil Sci 61:950–957. https://doi.org/10.1111/j.1365-2389.2010.01292.x

Todd-Brown KEO, Randerson JT, Post WM et al (2013) Causes of variation in soil carbon simulations from CMIP5 Earth system models and comparison with observations. Biogeosciences 10:1717–1736. https://doi.org/10.5194/bg-10-1717-2013

Toensmeier E, Ferguson R, Mehra M (2020) Perennial vegetables: a neglected resource for biodiversity, carbon sequestration, and nutrition. PLoS ONE 15(7):e0234611. https://doi.org/10.1371/journal.pone.0234611

Townsend-Small A, Czimczik CI (2010) Carbon sequestration and greenhouse gas emissions in urban turf. Geophys Res Lett 37:L02707. https://doi.org/10.1029/2009GL041675

Trammell TLE, Pataki DE, Pouyat RV et al (2020) Urban soil carbon and nitrogen converge at a continental scale. Ecol Monogr 90:e01401. https://doi.org/10.1002/ecm.1401

Trost B, Prochnow A, Drastig K et al (2013) Irrigation, soil organic carbon and N_2O emissions. A review. Agron Sustain Dev 33:733–749. https://doi.org/10.1007/s13593-013-0134-0

Tubiello FN, Conchedda G, Wanner N et al (2021) Carbon emissions and removals from forests: new estimates, 1990–2020. Earth Syst Sci Data 13:1681–1691. https://doi.org/10.5194/essd-13-1681-2021

Turetsky MR, Kane ES, Harden JW et al (2011) Recent acceleration of biomass burning and carbon losses in Alaskan forests and peatlands. Nat Geosci 4:27–31. https://doi.org/10.1038/ngeo1027

Ueyama M, Ichii K, Kobayashi H et al (2020) Inferring CO_2 fertilization effect based on global monitoring land-atmosphere exchange with a theoretical model. Environ Res Lett 15:084009. https://doi.org/10.1088/1748-9326/ab79e5

United States Department of Agriculture (2018) Summary report: 2015 national resources inventory. Natural Resources Conservation Service/Center for Survey Statistics and Methodology, Iowa State University, Washington, DC/Ames

United States Department of Agriculture (2019) U.S. census of agriculture, WP1905-2017, Census Report. USDA, Washington, DC

United States Department of Agriculture (2020) USDA National Agricultural 2020 Statistics service NASS – QuickStats 2020

United States Environmental Protection Agency (2021) Inventory of U.S. greenhouse gas emissions and sinks: 1990–2019. EPA 430-R-21-005. U.S. Environmental Protection Agency, Washington, DC. https://www.epa.gov/ghgemissions/inventory-us-greenhouse-gas-emissions-and-sinks

United States Geological Survey (2020) Mineral Commodity Summaries, January 2020. https://pubs.usgs.gov/periodicals/mcs2020/mcs2020-peat.pdf

Van Gestel N, Shi Z, van Groenigen KJ et al (2018) Predicting soil carbon loss with warming. Nature 554:E4–E6. https://doi.org/10.1038/nature25745

Van Mantgem PJ, Stephenson NL, Byrne JC et al (2009) Widespread increase of tree mortality rates in the western United States. Science 323:521–524. https://doi.org/10.1126/science.1165000

Varney RM, Chadburn SE, Friedlingstein P et al (2020) A spatial emergent constraint on the sensitivity of soil carbon turnover to global warming. Nat Commun 11:5544. https://doi.org/10.1038/s41467-020-19208-8

Verheyen K, Vanhellemont M, Auge H et al (2016) Contributions of a global network of tree diversity experiments to sustainable forest plantations. Ambio 45:29–41. https://doi.org/10.1007/s13280-015-0685-1

Verhoeven JTA, Setter TL (2010) Agricultural use of wetlands: opportunities and limitations. Ann Bot 105:155–163. https://doi.org/10.1093/aob/mcp172

Verkerk PJ, Costanza R, Hetemäki L et al (2020) Climate-smart forestry: the missing link. Forest Policy Econ 115:102164. https://doi.org/10.1016/j.forpol.2020.102164

Vesterdal L, Clarke N, Sigurdsson BD, Gundersen P (2013) Do tree species influence soil carbon stocks in temperate and boreal forests? For Ecol Manag 309:4–18. https://doi.org/10.1016/j.foreco.2013.01.017

Virkkala AM, Aalto J, Rogers BM et al (2021) Statistical upscaling of ecosystem CO_2 fluxes across the terrestrial tundra and boreal domain: regional patterns and uncertainties. Glob Chang Biol 27:4040–4059. https://doi.org/10.1111/gcb.15659

Vitharana UWA, Mishra U, Jastrow JD, Matamala R, Fan Z (2017) Observational needs for estimating Alaskan soil carbon stocks under current and future climate. J Geophys Res Biogeosci 122:415–429. https://doi.org/10.1002/2016JG003421

Vose JM, Peterson DL, Patel-Weynand T (eds) (2012) Effects of climatic variability and change on forest ecosystems: a comprehensive science synthesis for the U.S. forest sector. USDA, Forest Service, Pacific Northwest Research Station General Technical Report PNW-GTR-870

Walker X, Baltzer JL, Cumming SG et al (2019) Increasing wildfires threaten historic carbon sink of boreal forest soils. Nature 572:520–523. https://doi.org/10.1038/s41586-019-1474-y

Walker XJ, Rogers BM, Veraverbeke S et al (2020) Fuel availability not fire weather controls boreal wildfire severity and carbon emissions. Nat Clim Chang 10:1130–1136. https://doi.org/10.1038/s41558-020-00920-8

Walthall CJ, Hatfield J, Backlund et al (2013) Climate change and agriculture in the United States: effects and adaptation. U.S. Department of Agriculture, Washington, DC

Wang S, Huang Y (2020) Determinants of soil organic carbon sequestration and its contribution to ecosystem carbon sinks of planted forests. Glob Chang Biol 26:3163–3173. https://doi.org/10.1111/gcb.15036

Wang Y, Shipley BR, Lauer DA, Pineau RM, McGuire JL (2020a) Plant biomes demonstrate that landscape resilience today is the lowest it has been since end-Pleistocene megafaunal extinctions. Glob Chang Biol 26:5914–5927. https://doi.org/10.1111/gcb.15299

Wang S, Zhang Y, Ju W et al (2020b) Recent global decline of CO_2 fertilization effects on vegetation photosynthesis. Science 370:1295–1300. https://doi.org/10.1126/science.abb7772

Wang JA, Baccini A, Farina M, Randerson JT, Friedl MA (2021a) Disturbance suppresses the aboveground carbon sink in North American boreal forests. Nat Clim Chang. https://doi.org/10.1038/s41558-021-01027-4

Wang J, Defrenne C, McCormack ML et al (2021b) Fine-root functional trait responses to experimental warming: a global meta-analysis. New Phytol 230:1856–1867. https://doi.org/10.1111/nph.17279

Watts JD, Natali SM, Minions C et al (2021) Soil respiration strongly offsets carbon uptake in Alaska and Northwest Canada. Environ Res Lett 16:084051. https://doi.org/10.1088/1748-9326/ac1222

Wear DN, Coulston JW (2015) From sink to source: regional variation in U.S. forest carbon futures. Sci Rep 5(16518). https://doi.org/10.1038/srep16518

Wentland SC, Ancona ZH, Bagstad KJ et al (2020) Accounting for land in the United States: integrating physical land cover, land use, and monetary valuation. Ecosyst Serv 46:101178. https://doi.org/10.1016/j.ecoser.2020.101178

Wichtmann W, Joosten H (2007) Paludiculture: peat formation and renewable resources from rewetted peatlands. IMCG Newsl

Wieder WR, Hartman MD, Sulman BN et al (2018) Carbon cycle confidence and uncertainty: exploring variation among soil biogeochemical models. Glob Chang Biol 24:1563–1579. https://doi.org/10.1111/gcb.13979

Wildlands League (2019) Logging scars. https://loggingscars.ca/report/

Williams CA, Gu H, MacLean R, Masek JG, Collatz GJ (2016) Disturbance and the carbon balance of US forests: a quantitative review of impacts from harvests, fires, insects, and droughts. Glob Planet Chang 143:66–80. https://doi.org/10.1016/j.gloplacha.2016.06.002

Williams AP, Cook ER, Smerdon JE et al (2020a) Large contribution from anthropogenic warming to an emerging North American megadrought. Science 368:314–318. https://doi.org/10.1126/science.aaz9600

Williams MI, Farr CL, Page-Dumroese DS, Connolly SJ, Padley E (2020b) Soil management and restoration. In: Pouyat R, Page-Dumroese D, Patel-Weynand T, Geiser L (eds) Forest and rangeland soils of the United States under changing conditions. Springer, Cham, pp 145–167. https://doi.org/10.1007/978-3-030-45216-2_8

Williams CA, Gu H, Jiao T (2021) Climate impacts of U.S. forest loss span net warming to net cooling. Sci Adv 7:eaax8859. https://doi.org/10.1126/sciadv.aax8859

Wills S, Loecke T, Sequeira C et al (2014) Overview of the U.S. Rapid Carbon Assessment Project: sampling design, initial summary and uncertainty estimates. In: Hartemink A, McSweeney K (eds) Soil carbon. Progress in soil science. Springer, Cham, pp 95–104. https://doi.org/10.1007/978-3-319-04084-4_10

Wilsey B, Xu X, Polley HW, Hofmockel K, Hall SJ (2020) Lower soil carbon stocks in exotic *vs.* native grasslands are driven by carbonate losses. Ecology 101:e03039. https://doi.org/10.1002/ecy.3039

Wu D, Piao S, Zhu D et al (2020) Accelerated terrestrial ecosystem carbon turnover and its drivers. Glob Chang Biol 26:5052–5062. https://doi.org/10.1111/gcb.15224

Wyckoff WB (2015) Alaska farmer turns icy patch of Tundra into a breadbasket. https://www.npr.org/sections/thesalt/2015/02/26/389011370/alaska-farmer-turns-icy-patch-of-tundra-into-a-breadbasket

Xiao L, Zhou S, Zhao R, Greenwood P, Kuhn NJ (2020) Evaluating soil organic carbon stock changes induced by no-tillage based on fixed depth and equivalent soil mass approaches. Agric Ecosyst Environ 300:106982. https://doi.org/10.1016/j.agee.2020.106982

Xiong X, Grunwald S, Myers DB et al (2014) Interaction effects of climate and land use/land cover change on soil organic carbon sequestration. Sci Total Environ 493:974-982. http://dx.doi.org/https://doi.org/10.1016/j.scitotenv.2014.06.088

Xu S, Geng W, Sayer EJ et al (2020a) Soil microbial biomass and community responses to experimental precipitation change: a meta-analysis. Soil Ecol Lett 2:93–103. https://doi.org/10.1007/s42832-020-0033-7

Xu H, Xiao J, Zhang Z et al (2020b) Canopy photosynthetic capacity drives contrasting age dynamics of resource use efficiencies between mature temperate evergreen and deciduous forests. Glob Chang Biol 26:6156–6167. https://doi.org/10.1111/gcb.15312

Xu H, Vandecasteele B, De Neve S, Boeckx P, Sleutel S (2021a) Contribution of above- versus belowground C inputs of maize to soil organic carbon: conclusions from a $^{13}C/^{12}C$-resolved resampling campaign of Belgian croplands after two decades. Geoderma 383:114727. https://doi.org/10.1016/j.geoderma.2020.114727

Xu C, Xu X, Ju C et al (2021b) Long-term, amplified responses of soil organic carbon to nitrogen addition worldwide. Glob Chang Biol 27:1170–1180. https://doi.org/10.1111/gcb.15489

Yang P, Zhao Q, Cai X (2020) Machine learning based estimation of land productivity in the contiguous US using biophysical predictors. Environ Res Lett 15:074013. https://doi.org/10.1088/1748-9326/ab865f

Yates CF, Guo J, Bell TH et al (2021) Tree-induced alterations to soil properties and rhizoplane associated bacteria following 23 years in a common garden. Plant Soil 461:591–602. https://doi.org/10.1007/s11104-021-04846-8

Young MD, Ros GH, de Vries W (2021) Impacts of agronomic measures on crop, soil, and environmental indicators: a review and synthesis of meta-analysis. Agric Ecosyst Environ 319:107551. https://doi.org/10.1016/j.agee.2021.107551

Yu Z, Griffis TJ, Baker JM (2021a) Warming temperatures lead to reduced summer carbon sequestration in the U.S. Corn Belt. Nat Commun Earth Environ 2(53). https://doi.org/10.1038/s43247-021-00123-9

Yu C, Miao R, Khanna M (2021b) Maladaptation of U.S. corn and soybeans to a changing climate. Sci Adv 11:12351. https://doi.org/10.1038/s41598-021-91192-5

Yu W, Weintraub SR, Hall SJ (2021c) Climatic and geochemical controls on soil carbon at the continental scale: interactions and thresholds. Glob Biogeochem Cycles 35:e2020GB006781. https://doi.org/10.1029/2020GB006781

Yue C, Ciais P, Zhu D et al (2016) How have past fire disturbances contributed to the current carbon balance of boreal ecosystems? Biogeosciences 13:675–690. https://doi.org/10.5194/bg-13-675-2016

Zhang X, Guan D, Li W et al (2018) The effects of forest thinning on soil carbon stocks and dynamics: a meta-analysis. For Ecol Manag 429:36–43. https://doi.org/10.1016/j.foreco.2018.06.027

Zhang Y, Gurung R, Marx E et al (2020a) DayCent model predictions of NPP and grain yields for agricultural lands in the contiguous U.S. J Geophys Res Biogeosci:125, e2020JG005750. https://doi.org/10.1029/2020JG005750

Zhang Y, Marx E, Williams S et al (2020b) Adaptation in U.S. Corn Belt increases resistance to soil carbon loss with climate change. Sci Rep 10:13799. https://doi.org/10.1038/s41598-020-70819-z

Zhang X, Lark TJ, Clark CM, Yuan Y, LeDuc SD (2021) Grassland-to-cropland conversion increased soil, nutrient, and carbon losses in the US Midwest between 2008 and 2016. Environ Res Lett 16:054018. https://doi.org/10.1088/1748-9326/abecbe

Zheng W, Holm N, Spokas KA (2015) Research and application of biochar in North America. In: Guo M, He Z, Uchimiya M (eds) Agricultural and environmental applications of biochar: ADVANCES and barriers. SSSA Special Publication 63. Madison, pp 1–20. https://doi.org/10.2136/sssaspecpub63.2014.0053

Zhou YT, Gowda PH, Wagle P et al (2019a) Climate effects on tallgrass prairie responses to continuous and rotational grazing. Agronomy 9:219. https://doi.org/10.3390/agronomy9050219

Zhou G, Xu S, Ciais P et al (2019b) Climate and litter C/N ratio constrain soil organic carbon accumulation. Nat Sci Rev 6:746–757. https://doi.org/10.1093/nsr/nwz045

Zhou L, Hong Y, Li C et al (2020) Responses of biomass allocation to multi-factor global change: a global synthesis. Agric Ecosyst Environ 304:107115. https://doi.org/10.1016/j.agee.2020.107115

Zhu Z, McGuire AD (eds) (2016) Baseline and projected future carbon storage and greenhouse-gas fluxes in ecosystems of Alaska. U.S. Geological Survey Professional Paper 1826, 196 p. https://doi.org/10.3133/pp1826

Zomer RJ, Bossio DA, Sommer R, Verchot LV (2017) Global sequestration potential of in-creased organic carbon in cropland soils. Sci Rep 7:15554. https://doi.org/10.1038/s41598-017-15794-8

Chapter 4
Soil Inorganic Carbon Stocks in Terrestrial Biomes

Abstract About 40% of the soil carbon (C) stock to 200-cm (79-in) soil depth in the conterminous U.S. (CONUS, i.e., the 48 adjoining U.S. states on the continent of North America) is comprised of soil inorganic carbon (SIC), estimated at 54.1 Pg C (59.6 million tn C). Soil and land-use management practices affect SIC with croplands and shrublands having the highest SIC stocks. The SIC consists of carbonates and bicarbonates, further differentiated as primary or lithogenic carbonates inherited from calcareous soil parent material or newly formed secondary or pedogenic carbonates. The formation of the latter can be an effective soil C sequestration mechanism when calcium (Ca) precipitated with pedogenic carbonates is released directly from silicates resulting in the formation of silicatic pedogenic carbonates. Unlike SOC which is controlled primarily by the vegetative community, SIC stocks can continue to accumulate over hundreds of thousands of years to highly concentrated levels, such as those in petrocalcic horizons. Thus, the changes in the SIC stock must also be considered in a credible assessment of soil C sequestration in terrestrial biomes of the U.S. aside that of changes in the SOC stock. Carbonate rocks are pure limestone, chalk, dolomitic limestone, dolostone and dolomite, and are found on 15% of the non-glaciated area of North America. Calcareous bedrocks are widely distributed in the U.S. Formation and accumulation of secondary carbonates (calcification) occurs in about 10% of U.S. soil area. Formation of SIC in terrestrial biomes of the U.S. can be enhanced by: (i) irrigation management, (ii) additions of dolomite, gypsum or lime, and magnesium (Mg) with rock minerals, and (iii) application of organic and mineral amendments resulting in higher soil inputs of Ca^{2+}, Mg^{2+} and/or sodium (Na^{+}). However, soil analysis does not routinely distinguish between lithogenic from pedogenic carbonates. Thus, the contribution of SIC formation to soil C sequestration in terrestrial biomes of the U.S. is a researchable priority. This chapter begins with an overview on processes resulting in SIC sequestration. Then, evidence for the occurrence of primary or lithogenic and for secondary or pedogenic carbonates in soils of the U.S. is presented. The chapter concludes with a discussion on how SIC stocks can potentially be managed.

K. Lorenz, R. Lal, *Soil Organic Carbon Sequestration in Terrestrial Biomes of the United States*, https://doi.org/10.1007/978-3-030-95193-1_4

Keywords Soil inorganic carbon · Primary carbonates · Lithogenic carbonates · Secondary carbonates · Pedogenic carbonates · Irrecoverable carbon · Land Resource Regions · Major Land Resource Areas · Petrocalcic horizon · Soil and Land-Use Management

4.1 Introduction

The assessment and management of soil carbon (C) sequestration has mainly focused on the soil organic carbon (SOC) stock. However, the other principal soil C form, i.e., carbonate-C, inorganic C or soil inorganic carbon (SIC) must also be considered. Gaseous carbon dioxide (CO_2)(g), dissolved CO_2(aq), carbonic acid H_2CO_3(aq), bicarbonate HCO_3^-(aq), carbonate CO_3^{2-}(aq), and solid-phase calcium carbonate (mainly calcite; $CaCO_3$) are collectively referred to as inorganic C (Monger 2014). The SIC stock consists mainly of carbonates and bicarbonates further differentiated as primary or lithogenic carbonates inherited from calcareous soil parent material or newly formed secondary or pedogenic carbonates (Monger et al. 2015). Zamanian et al. (2016) differentiated also biogenic carbonates formed within terrestrial or aquatic animals and plants as part of their skeleton. This includes for example shells, bones and calcified seeds, or released from or within certain organs such as the esophageal glands of earthworms. Importantly, Zamanian and Kuzyakov (2019) suggested that the contribution of SIC to atmospheric CO_2 may be more important than previously thought which also refers to its potential role for C sequestration (Monger 2014; Schlesinger 2000). In addition to C sequestration, carbonates may also affect SOC accumulation and turnover by acting as cementing agents which is relevant for soil aggregation (Totsche et al. 2018; Oades and Waters 1991). Further, mineral carbonates, through major changes of surface acidity at the mineral-water interface, play an important role in the hydrolytic breakdown of macromolecular organic molecules in soil (Kleber et al. 2021).

This chapter begins with an overview on processes resulting in SIC sequestration. Then, evidence for the occurrence of primary or lithogenic and for secondary or pedogenic carbonates in soils of the U.S. is presented. The chapter concludes with a discussion on how SIC stocks can potentially be managed.

4.1.1 The Process of Soil Inorganic Carbon Sequestration

Sequestration by SIC formation depends on the calcium (Ca) source as no CO_2 is sequestered if Ca is derived from pre-existing carbonates during the formation of calcitic pedogenic carbonates (Monger et al. 2015). However, the formation of pedogenic carbonates can be an effective C sequestration mechanism when Ca^{2+} precipitated with pedogenic carbonates is released directly from silicates resulting in the formation of silicatic pedogenic carbonates (Monger et al. 2015). Specifically, of the two molecules CO_2 taken up during pedogenic carbonate formation (Eq. 4.1), one molecule of CO_2 is sequestered (Eq. 4.2):

$$CaSiO_3 + 3H_2O + 2CO_2 \rightarrow Ca^{2+} + 2HCO_3^- + H_4SiO_4 \tag{4.1}$$

$$Ca^{2+} + 2HCO_3^- \rightarrow CaCO_3 \downarrow + H_2O + CO_2 \tag{4.2}$$

Maintaining and/or accumulating the SIC stock is not among the aims of sustainable soil and land-use management practices such as conservation agriculture despite its effects on soil processes. For example, carbonates influence soil microbial activity, the rate of SOC mineralization, soil pH, adsorption-desorption processes, and soil cementation (Garcia et al. 2018). The presence of $CaCO_3$ in soil, in particular, is important for acidity buffering, aggregate formation and stabilization, SOC stabilization, microbial and enzyme activities, nutrient cycling and availability, and water permeability and plant productivity (Raza et al. 2020). Calcium has also been linked to increased organo-mineral association in soils (Rowley et al. 2021), which contributes to SOC sequestration.

4.1.2 Changes in Soil Inorganic Carbon Stocks

There is little to mixed evidence of land-use effects on SIC stocks but data are often incomplete due to shallow sampling depths of many studies and inventories (An et al. 2019). For example, Kim et al. (2020) compiled more than one hundred comparisons of SIC stocks under natural vegetation and croplands with an even split of comparisons reporting higher SIC stocks with agricultural conversions. However, most of the studies focused on the surface soil (sampling depth of <2.5 m [8.2 ft] with median depth of 0.75 m [2.5 ft]). Changes in environmental properties such as soil acidification induced by nitrogen (N) fertilization, N fixation by legumes or intensification of rewetting cycles due to irrigation but also in atmospheric deposition may release great amounts of SIC (Zamanian et al. 2016). Losses of SIC by HCO_3^- leaching and CO_2 efflux from calcareous soils alone have been estimated at 7.5 Tg C year^{-1} (8.3 million tn C year^{-1}; Raza et al. 2021). In the 1980s–2010s, SIC stocks in 0–30 cm depth in China decreased by 8.99% (1.37 Pg C [1.51 billion tn C]; Song et al. 2021). This loss was equivalent to 17.6–24.0% of the terrestrial C sink in China, and 57.1% of the cropland SOC sink in China. Thus, SIC losses may offset a large portion of the global efforts aimed at ecosystem C sequestration (Song et al. 2021). The acidification-induced SIC-borne CO_2 losses, in particular, may be a major soil C loss pathway, and jeopardize global efforts to mitigate climate change through SOC sequestration (Raza et al. 2021).

Importantly, all released SIC is irrecoverable because, after leaching of Ca^{2+} and Mg^{2+} (with accompanying nitrate), no other cations are available to trap the CO_2 released by root respiration and the microbial decomposition of organic matter (OM; Zamanian et al. 2021). Over the last 50 years, at least 0.41 Pg C (0.45 billion tn C) may have been released irrecoverably as CO_2 to the atmosphere from the agricultural SIC stock. Erosion, weathering and dissolution have been proposed as mechanisms for SIC loss by land use. For example, deepening roots in woodlands can enhance carbonate weathering by promoting recharge and CO_2–carbonate contact in the deep subsoil, with the hydrological impacts of rooting characteristics more influential than those of soil CO_2 distribution in modulating carbonate weathering rates (Wen et al. 2021). Otherwise, carbonate precipitation, i.e., by addition of Ca-rich irrigation waters or enhanced freeze–thaw cycles have been hypothesized as possible causes for increases in SIC stocks. Importantly, changes in SIC stocks can be sometimes more than one order of magnitude larger than those in SOC stocks (Kim et al. 2020). Overall, SIC is not stable as previously thought, and can also be employed by certain microbes and transformed into SOC (Liu et al. 2021). Specifically, soil microorganisms in drylands may employ the chemoautotrophic and heterotrophic pathways, namely, dark microbial pathways, for SIC transformation into SOC (Liu et al. 2021).

Globally, the SIC stock contains an estimated 947 Pg C (1044 billion tn C), i.e., more than double the amount of plant biomass C (450 Pg C [496 billion tn C]; Eswaran et al. 2000; Bar-On et al. 2018). The highest SIC stocks – 320 to 1280 Mg C ha^{-1} (143–571 tn C ac^{-1}) – are accumulated in soils of arid regions with mean annual precipitation (MAP) below 250 mm (9.8 in; Zamanian et al. 2016). However, there is also a large SIC stock outside of arid and semi-arid regions (Guo et al. 2006a). For example, Guo et al. (2006b) performed the first estimations of SIC stocks of the conterminous U.S. (CONUS, i.e., the 48 adjoining U.S. states on the continent of North America), an area of 7.37 million km^2 (2.85 million mi^2) excluding urban land, bare rock, and other non-soil bodies (Table 4.1). In the central U.S., a substantial SIC stock exists between depths of 1 and 2 m (39.4 and 78.8 in). Overall, the SIC stock may amount to about 40% to the total soil C stock in 0–200 cm (0–78.8 in) depth in CONUS (Guo et al. 2006b). There was no linear relationship between SOC and SIC at any soil depth as SOC is primarily derived from plants and soil biota while SIC originates partially from both soil parent material and biological sources (Monger 2014; Guo et al. 2006a, b).

Terrestrial biomes and ecosystems in CONUS may differ not only in SOC stocks but also in SIC stocks. About 22% of the total SIC stock was in the Land Resource Regions (LRRs) Western Range and Irrigated Region, 19% in the Central Great Plains Winter Wheat and Range Region, and 11% in the Central Feed Grains and Livestock Region (Guo et al. 2006a). Similar to SOC stocks, human disturbance has reduced SIC stocks in the surface layer in CONUS (Guo et al. 2006b). There was no obvious pattern in SIC stock as MAP increases until MAP exceeds 1000 mm (39.4 in), at which point on SIC drops dramatically.

To sum up, the SIC stock and its dynamic must also be considered in credible assessment of soil C sequestration in terrestrial biomes of the U.S. aside that of the SOC stock. Unlike SOC which is controlled primarily by the vegetative community, SIC continues to accumulate over hundreds of thousands of years to highly concentrated levels, such as those in petrocalcic horizons (Monger 2014). Thus, the net effects of soil cultivation and the possibility of managing SIC sequestration and

Table 4.1 Soil inorganic carbon stocks of terrestrial ecosystems in the conterminous United States (Guo et al. 2006b)

Soil depth		Terrestrial ecosystem	Soil inorganic carbon stock	
cm	in		Pg C	Billion tn C
0–20	0–8	Agriculture	0.54	0.60
		Forest	0.43	0.47
		Grass	1.13	1.25
		Pasture	0.23	0.25
		Shrub	1.88	2.07
		Wetland	0.07	0.77
		Total	4.13	4.55
0–100	0–39	Agriculture	6.44	7.10
		Forest	2.44	2.69
		Grass	7.26	8.00
		Pasture	2.07	2.28
		Shrub	9.25	10.20
		Wetland	0.62	0.68
		Total	28.14	31.02
0–200	0–79	Agriculture	15.09	16.63
		Forest	4.55	5.02
		Grass	12.89	14.21
		Pasture	4.63	5.10
		Shrub	15.40	16.98
		Wetland	1.23	1.36
		Total	54.14	59.68

stocks for soil C sequestration and increases in soil C stocks for climate change adaptation and mitigation, and for improving soil properties and health are researchable priorities.

4.2 Primary or Lithogenic Carbonates

Primary or lithogenic carbonates may contribute to the SIC stock in U.S. soils. Primary or lithogenic carbonates have remained or are inherited from soil parent materials or bedrock such as limestone particles or allocated onto the soil from other locations by calcareous dust or landslides (Zamanian et al. 2016). Further differentiated can be *in situ* lithogenic carbonates derived from residual limestone fragments above bedrock (Monger et al. 2015). Otherwise, *ex situ* lithogenic carbonate indicates transport from other locations, and may be divided into eolian, alluvial, colluvial, or glacial. Carbonate rocks are pure limestone, chalk, dolomitic limestone, dolostone and dolomite (Fig. 4.1; Dürr et al. 2005). About 15% of the surficial lithology of the non-glaciated area of North America consists of carbonate rocks. Thus, carbonate rocks may contribute to the SIC stock of terrestrial biomes in the

Fig. 4.1 Put-in-Bay Dolomite over Tymochtee Dolomite (Upper Silurian; South Bass Island, Lake Erie, Ohio, USA; Creative Commons Attribution 2.0 Generic license)

U.S. For example, the SIC pattern in the south central plains (particularly Texas) matches the pattern of bedrock (Guo et al. 2006a), and the amount of SIC is partially determined by the soil parent material (Guo et al. 2006b). However, routine lab analysis cannot distinguish between lithogenic and pedogenic carbonates (Monger et al. 2015).

A large SIC stock was found to 2-m (78.7-in) depth in the U.S. Midwest where MAP ranges from 700 to 1000 mm (27.6 to 39.4 in; Guo et al. 2006b). While the SIC in the upper 1.0 m (39.4 in) is generally leached out in these climates, the deeper depth increment still retains some combination of primary/lithogenic and secondary/pedogenic carbonates. The presence of deep carbonate is particularly well correlated with the presence of young glacial deposits which still contain a significant pool of lithogenic carbonate (Guo et al. 2006a). Most of the agriculture (row crops) in the Midwest and Northern plains regions is located on soils originating from calcareous latest Pleistocene/early Holocene glacial geological substrates. This partially explains the high SIC content that remains in the deeper layers even though the high MAP has contributed to the removal of most of the carbonate from the upper 1 m (39.4 in; Guo et al. 2006b). However, data on lithogenic C stocks for soils of the U.S. are not available. In lieu, examples for LRRs and Major Land Resource Areas (MLRAs) in the U.S. where limestone and other carbonate-containing rocks are found are presented in the following section.

The LLR Central Great Plains Winter Wheat and Range Region is centered over western Kansas, and extends from the Nebraska Sand Hills to northern Texas

(Bockheim 2017). The region includes the Rolling Limestone Prairie MLRA with grassland as dominant vegetation. Residuum from limestone is among the soil parent materials. Limestone is also one of the dominant parent materials in the Southwest Plateaus and Plains Range and Cotton Region MLRAs. This region is limited to southern Texas, and under short-grass prairie, with oak savanna and desert shrubs in some areas. Further, residuum from limestone is common parent material in the Cherokee Prairies MLRA where the vegetation is mainly grasslands. The prairies are located in the LLR Central Feed Grains and Livestock Region. The High Limestone Plateaus MLRA are part of the Mariana Islands (Guam and the Commonwealth of the Northern Mariana Islands). Here, limestone residuum is among the soil parent materials. The vegetation is tropical forest, strand forests (linear swamp forests), and tropical palms, with abundant grasslands (Bockheim 2017).

Limestones are found in the Blue and Seven Devils Mountains MLRA in northeastern Oregon covered by forests and grasslands (McDaniel 2017). Extensive areas of limestone occur on the Colorado Plateau MLRA (Rasmussen et al. 2017). Three quarters of this area is rangeland used for sheep (*Ovis aries* Linnaeus, 1758) and cattle (*Bos bos taurus* Linnaeus, 1758) production. Another land use is juniper (*Juniperus* spp.) and pinyon–juniper woodland that are used for firewood and pinyon nut production in addition to grazing. Small areas of cropland are irrigated, and used to grow alfalfa (*Medicago sativa* L.), hay, and corn (*Zea mays* L.) on the Colorado Plateau. Areas of the Kaibab Plateau surrounding the Grand Canyon are also underlain by limestone. Dolostone is among the soil parent materials in the Mogollon Transition, and Arizona and New Mexico Mountains MLRAs covered by ponderosa pine (*Pinus ponderosa* Douglas ex C.Lawson; Rasmussen et al. 2017).

The Badlands of northwestern Nebraska and western South Dakota MLRA is a product of differential erosion of bedrock, including that of non-marine limestones (Hirmas and Mandel 2017). Soils of the Great Plains developed in glacial till commonly contain fragments of weathered limestone, but they generally comprise less than 50% of the >2 mm sediment fraction. Some of the area is under dryland farmland and rangeland. Cultivated areas are used for growing wheat (*Triticum aestivum* L.), barley (*Hordeum vulgare* L.), oats (*Avena sativa* L. (1753)), corn, alfalfa, feed grains, flax (*Linum usitatissimum* L.), hay and sunflowers (*Helianthus* L.). In the Flint Hills of Kansas, footslope positions are composed of a mix of fine- and coarse-grained colluvium derived from the upslope hard limestone bedrock. Limestone gravels or chert can reach up to 35% by volume in these soils. The soils are commonly used for rangeland or timber production (Hirmas and Mandel 2017).

Limestone parent materials are common in the Grand Prairie MLRA spanning Texas and Oklahoma (Ruppert 2017). Limestones form deposits more than a mile deep in the Arbuckle Mountains. The Texas Central Basin, Edwards Plateau and Rio Grande Plains MLRAs also contain limestones. Mixed oak savannah with mid/tall-grasses are found in the Texas Central Basin MLRA while rangeland predominates as the major land use at the Edwards Plateau. Grassland, shrubs and trees cover the Rio Grande Plains (Ruppert 2017).

The eastern Upper Peninsula of Michigan and parts of western Wisconsin and eastern Minnesota contain carbonate rocks like limestone and dolomite (Schaetzl 2017). Most of the landscape is forested as many soils are too wet, or too dry and sandy to economically cultivate for agriculture. Other soils are too far north and, thus, too cold for the production of most cash grains (Schaetzl 2017).

Limestone is found in the Till Prairies and Loess Hills MLRAs which are part of the Central Lowland Province of the Interior Plains (Libohova et al. 2017). Various tree species occur but almost the entire area is farmland used for the production of corn, soybeans (*Glycine max* L.), other feed grains, and hay. The Deep Loess and Drift MLRAs, and Iowa and Missouri Heavy Till Plain MLRA have limestone among the dominant geological deposits. More than 90% of the area is farmland with livestock production including beef cattle and swine (*Sus scrofa domesticus* Erxleben, 1777). The Till Plain MLRAs in the Central Lowland Province of the Interior Plains are underlain by limestone bedrock and dolostone. More than 75% of the region is farmland, and dairy farms and truck and canning crops are also widespread (Libohova et al. 2017).

The Central Claypan Areas MLRA is underlain by limestone bedrock, and supports tall prairie grasses, mainly big bluestem (*Andropogon gerardii* Vitman), Indiangrass (*Sorghastrum nutans* (L.) Nash), prairie dropseed (*Sporobolus heterolepis* A.Gray), and switchgrass (*Panicum virgatum* L.; Libohova et al. 2017). Further, about 70% of the area is in cropland use with the dominant crops corn, soybeans, other feed grains, and hay for cattle and other livestock. Other grains such as winter wheat, oats, and sorghum (*Sorghum bicolor* (L.) Moench) are also grown. Limestone and dolomite dominate rocks of the Western Michigan Fruit Belt MLRA, Southwestern Michigan Fruit and Truck Crop Belt MLRA, Southern Michigan and Northern Indiana Drift Plain MLRA, Erie-Huron Lake Plain MLRA, Ontario-Erie Plain and Finger Lakes Region MLRA. Broadleaf deciduous forest comprised of a wide range of species, including white oak (*Quercus alba* L.), black oak (*Quercus velutina* Lam), red maple (*Acer rubrum* L. 1753), sugar maple (*Acer saccharum* Marshall), and beech (*Fagus* spp.) cover the Western Michigan Fruit Belt MLRA. The forest extent decreases further south, where more than 75% of the land is farmed. The orchards, vegetables, crops, and dairy farms that dominate the northern part of the region are gradually replaced with corn, soybean, winter wheat, canning crops (such as sugar beets [*Beta vulgaris*]), and fruit and truck crops in areas of sandy soils (Libohova et al. 2017).

Most of the LRRs East and Central Farming and Forest Region, and Atlantic Basin Diversified Farming Region are underlain by sedimentary rocks including dolostone and limestone (Lee and Kabrick 2017). In the Interior Highlands, 50% of the land is forested, with forest cover ranging from 22% in the Springfield Plateau to 72% in the Ouachita Mountains. About one-third of the land is used for forages for beef and dairy industry while 5% is under urban development. There is a small amount of cropland (6%) primarily growing soybeans and winter wheat. This area contains one of the largest poultry production regions of the U.S. in northwestern Arkansas, northeastern Oklahoma, and southwestern Missouri. The Highland Rim and Pennyroyal Plateau MLRA, Nashville Basin MLRA and Kentucky Bluegrass

MLRA in the Interior Plains are characterized by limestone sinks and abundant outcrops of limestone. Much of the Interior Plains is in row crop production (27%) of corn, soybeans, wheat, or tobacco (*Nicotiana tabacum* L.) on the plateaus and other gently sloping terrains as well as on recently drained floodplains and bottomlands. Forests occupy 36% of the area, and commonly occur in fragmented woodlots on slopes too steep to farm or soils too shallow for row crops. Managed grasslands occupy 23% of the landscape, and are found on rolling terrain where soils are shallow. These areas are used for hay production or animal grazing. Urban areas occupy 7% of the landscape. Limestone bedrock occurs also in the Appalachian Highlands MLRAs. Land use there is predominantly forest (59%), while cropland and managed grasslands make up 26% of the land area (Lee and Kabrick 2017).

Some soils in the LRRs Mississippi Delta Cotton and Feed Grains Region; South Atlantic and Gulf Slope Cash Crops, Forest, and Livestock Region; and Atlantic and Gulf Coast Lowland Forest and Crop Region developed on calcareous parent materials (West et al. 2017). For example, many soils in the Alabama and Mississippi Blackland Prairie and Cretaceous Western Coastal Plain MLRAs are calcareous in lower horizons, and soils shallow to calcareous parent materials have high pH and carbonates throughout. Carbonates are also common in deeper soils horizons in the Arkansas and Red River Alluvium MLRAs. Soils in the LRR Mississippi Delta Cotton and Feed Grains Region are fertile and agriculturally productive. Crops, forests, and pasture are dominant land covers across the Southern and Western Coastal Plains. Forest production, pasture and production of row crops occurs in the LRR Atlantic and Gulf Coast Lowland Forest and Crop Region (West et al. 2017).

Within the LLR Northeastern Forage and Forest Region, limestone is sometimes incorporated in the till at the Tughill Plateau MLRA (Wilson and Shaw 2017). Land use is forests, with principal products of pulpwood, Christmas trees, and maple syrup. Limited acreage of pasture and cropland producing forage and small grains for dairy cattle are also present. Some soils in the St. Lawrence-Champlain Plain MLRA are derived from limestone/dolomite. Many of these soils are cleared and used for hay, pasture, corn, or small grains due to their more fertile nature. Some of the bedrock of New England and Eastern New York Upland, Northern Part MLRA are limestone and dolomite. Boreal conifer forests occupy higher elevations on mountain summits, while a combination of northern deciduous hardwoods and spruce-fir forests occupy middle slopes. White pine (*Pinus strobus* L.) and eastern hemlock (*Tsuga canadensis* (L.) Carrière) are other common species. There are some dairy and beef farms, and corn, potatoes (*Solanum tuberosum* L.), apples (*Malus domestica* Borkh., 1803), and beans are main cash crops. Dolomite and limestone or limestone sometimes dominate the rock types in the New England and Eastern New York Upland, Southern Part MLRA. The area is principally being central hardwood forest, with maples (*Acer rubrum* L. 1753, *A. saccharum* Marshall), birch (*Betula aleghaniensis* Britt.), and eastern hemlock (*Tsuga canadensis* (L.) Carrière) as associated forest components. Many rural areas are used for cultivated crops, hay, or pasture. Dairies, truck crops, apple, and nursery stock production are also common. The Aroostook Area MLRA is also underlain by limestone bedrock. Land use is principally forest, and native vegetation is a mixture of coniferous and deciduous trees (sugar maple, American beech (*Fagus grandifolia* Ehrh.), birch, quaking aspen (*Populus tremuloides* Michx.), red spruce (*Picea rubens* Sarg.), balsam fir (*Abies balsamea* (L.) Mill.), and white pine. The principal agricultural crop is potatoes, as well as some broccoli (*Brassica oleracea* L.), oats, canola (*Brassica napus* L.), and barley (Wilson and Shaw 2017).

Limestone bedrock is found in the LRR Subtropical Fruit, Truck Crop, and Range Region (Collins 2017). Native vegetation in this LRR includes live oak (*Quercus virginiana* Mill.), turkey oak (*Quercus laevis* Walter), slash pine (*Pinus elliottii* Engelm.), longleaf pine (*Pinus palustris* Mill.), Sable palmetto maiden cane (*Panicum hemitomon* Schult.), and various types of mangroves and grasses. Agricultural crops include winter vegetables, truck crops, citrus (*Citrus* L.), forest-land, and grazing lands for livestock. Soils developed from limestone occur in the South-Central Florida Ridge MLRA. Some of the soils are used for improved pasture or for various cultivated crops (e.g., peanuts (*Arachis hypogaea* L.), watermelon (*Citrullus lanatus* (Thunb.) Matsum. & Nakai), corn, and hay grasses). Native vegetation includes live oak, laurel oak (*Quercus laurifolia* Michx), and post oak (*Quercus stellate* Wangenh.). The dominant bedrock of the Florida Everglades and Associated Areas MLRA is also limestone. Most of the native vegetation is wet grasses [sawgrass (*Cladium mariscus* (L.) Pohl), pickleweed (*Salicornia* L.), buttonbush (*Cephalanthus occidentalis* L., 1753), and maiden cane]. Bald cypress (*Taxodium distichum* (L.) Rich.) is common in swamps, and mangrove trees are common in saltwater areas. A limited area of cropland is used to grow winter vegetables and citrus while some sugarcane (*Saccharum officinarum* L.) is also grown (Collins 2017).

In the LRR Caribbean Region, limestone is among the soil parent materials (Collins 2017). Food crops such as plantains (*Musa* × *paradisiaca*), bananas (*Musa*), yams (*Dioscorea* L.), vegetables, and some citrus fruits are grown as main crops. The Semi-arid Mountains and Valleys MLRA is about 70% in Puerto Rico, 20% in the U.S. Virgin Islands, and 10% on the outlying small islands. Soil parent materials consist of a mixture of limestone and volcanic rock. Vegetation is comprised of grasses (hurricane grass (*Fimbristylis cymose* R. Br.), guinea grass (*Megathyrsus maximus* (Jacq.) B.K. Simon & S.W.L. Jacobs, 2003), buffelgrass (*Cenchrus ciliaris* L.), and Egyptian grass (*Dactyloctenium aegyptium* (L.) Willd.) and trees (black olive (*Olea europaea* L.), turpentine (*Syncarpia glomulifera* (Sm.) Nied.), Christmas tree and guayacan (*Tabebuia chrysantha* (Jacq.) S.O.Grose)). Almost 40% of the MLRA is in grassland while private forested areas cover about 25%. The Humid Coastal Plains MLRA on the north coast of Puerto Rico is characterized by irregular karst landscapes associated with limestone. The vegetation in dry areas is a diverse suite of grasses and trees. In areas of poorly or very poorly drained soils, the vegetation consists mainly of mangroves, as well as southern cattail (*Typha domingensis* Pers.), leatherfern (*Acrostichum aureum* L.), and para grass (*Urochloa mutica* (Forssk.) Stapf). The Semi-arid Coastal Plain MLRA is principally located on the southern coast of Puerto Rico. Soil parent materials at higher elevations are limestone and volcanic rock. The vegetation across the region is diverse. In drier areas, plant species includes beachgrass (*Ammophila* Host), southern sandbur (*Cenchrus echinatus* L.), saltwort (*Batis maritime* L.), bermudagrass (*Cynodon dactylon* (L.) Pers.), flame tree, white oak, *Leucaena* Benth., and black olive. In wetter areas, the vegetation includes mangroves, southern cattail, leatherfern, water panicum (*Panicum* L.), and para grass. Approximately 50% of the land use is grassland and urban development while cropland comprises 20% of the area (Collins 2017).

Soils formed on coralline limestone plateaus on the Hawaiian islands Maui, Oahu, and Kauai may contain lithogenic carbonates (Robotham et al. 2017). Intensive agriculture is common with crops like sugarcane, corn and a wide variety of vegetables.

In the LRRs Interior and Western Alaska, some soils in the Alpine Zone developed on calcareous sedimentary rocks like shale or limestone (Ping et al. 2017). Vegetation is alpine shrub tundra. Further, calcareous fens occur in areas with limestone bedrock in the Southeast and Coastal Gulf of Alaska MLRA (Ping et al. 2017).

The previous section highlighted that calcareous bedrock is widely distributed in the U.S., and may contribute to primary or lithogenic soil carbonate stocks. For a more credible assessment of soil C sequestration and its management in terrestrial biomes of the U.S., the natural dynamic of primary or lithogenic carbonates, their contribution to the SIC stock, and the effects of management practices must be studied in detail.

4.3 Secondary or Pedogenic Carbonates

The formation of secondary or pedogenic carbonates (Eq. 4.2) can be an effective C sequestration mechanism in soil. Soils of arid, semiarid, and subhumid regions, in particular, are sinks for inorganic C in the form of pedogenic carbonates (Monger 2014). Secondary or pedogenic carbonates are formed and redistributed in soils via: (i) dissolution of SIC (i.e., primary, biogenic or previously formed pedogenic carbonates), (ii) movement of dissolved ions within pores, through soil profiles as well as landscapes, and (iii) re-precipitation of dissolved ions in various morphologies such as carbonate nodules (Zamanian et al. 2016). Distinct morphological features of pedogenic carbonates such as nodules and coatings form in various time spans—from a few weeks (e.g., hypocoatings) and decades (e.g., rhizoliths) to hundreds of thousands or even millions of years (e.g., calcrete, caliche or cemented horizon; Fig. 4.2; Zamanian et al. 2016).

Four conditions must be met that pedogenic carbonate formation occurs in soil (Monger 2014): (i) the soil must have an alkaline pH; (ii) there must be an active source of CO_2, i.e., root, microbial or faunal respiration, which is necessary for HCO_3^- production; (iii) there must be available Ca^{2+}; and (iv) the soil must occasionally contain moisture but not too much moisture as the formation of $CaCO_3$ is an aqueous phenomenon. Thus, the driest zones of the Atacama Desert in Chile are too dry for pedogenic carbonate formation. Otherwise, when rainfall increases above 500–760 mm (19.7–29.9 in), pedogenic carbonates do also not typically form in soil profiles. To sum up, promoting SIC sequestration by pedogenic carbonate

Fig. 4.2 Caliche berm surrounding a stock tank in Central Texas, USA (Creative Commons Attribution-Share Alike 3.0)

formation requires the manipulation of Ca supply from silicates, water, alkaline pH, and CO_2 (i.e., root, microbial or faunal respiration). Because soil bacteria and fungi are agents of calcite biomineralization, they may also be managed to enhance SIC sequestration by pedogenic carbonate formation (Monger 2014). Calcite is also more soluble at low temperatures. Thus, all else being equal, calcite is more likely to precipitate and be present in soils with high temperatures (Graham and Indorante 2017).

The formation and accumulation of secondary carbonates (calcification) occurs in 10% of U.S. soils, especially those in the LRRs Northwestern Wheat and Range Region, Western Range and Irrigated Region, and Rocky Mountain Range and Forest Region (Bockheim 2017). Translocation and precipitation of $CaCO_3$ can result in the formation of identifiable and visible secondary carbonates such as soft masses, nodules, coatings, and filaments which are diagnostic characteristics commonly observed in U.S. soils (Ditzler 2017). In addition, a distinct subsoil zone of calcite accumulation may form known as calcic horizon. If calcification occurs over hundreds of thousands of years, the horizon becomes cemented by calcite so that it is rock hard – a petrocalcic horizon has formed (Graham and Indorante 2017).

Some soils in desert or semidesert areas have a subsoil horizon into which significant amounts of $CaCO_3$ have accumulated. Water from precipitation dissolves $CaCO_3$ in the surface horizon and moves it into the subsoil. The water may be subsequently taken up by plant roots or may evaporate, leaving any dissolved minerals such as $CaCO_3$ behind. This diagnostic subsoil horizon is a calcic horizon (Ditzler

2017). Both calcic and petrocalcic horizons occur typically in arid or semiarid environments in the U.S. For example, in the LRR California Subtropical Fruit, Truck, and Specialty Crop Region, soils in the San Joaquin Valley may exhibit petrocalcic horizons (Southard et al. 2017). Throughout the valley, many of the soils with petrocalcic horizons have been deeply subsoiled in order to increase rooting depth and irrigation water penetration for agricultural land use. To increase irrigation water efficiency, thousands of hectares of soils have also been precision-levelled throughout the valley for the production of a wide variety of crops. Higher terrace positions have a rolling topography and are usually irrigated by drip or micro-sprinkler methods (Southard et al. 2017).

In the Palouse and Nez Perce Prairies MLRA, calcic diagnostic horizons are present in soils of the areas with lower MAP, whereas $CaCO_3$ is typically absent from the upper meter (39.4 in) of soils from the higher MAP areas (McDaniel 2017). On the drier, warmer western edge of the MLRA, big sagebrush (*Artemisia tridentata* Nutt.), Idaho fescue (*Festuca idahoensis* Elmer), and bluebunch wheatgrass (*Pseudoroegneria spicata* (Pursh) Á.Löve 1980) dominate. Little leaching has occurred in soils of the Columbia Basin and Columbia Plateau MLRAs, and common are horizons in which $CaCO_3$ has accumulated. In contrast, soils of the Northern Rocky Mountains, Valleys, and Foothills MLRAs leach $CaCO_3$ to greater soil depth as production increases along a moisture gradient in response to greater precipitation. The land use is mainly coniferous forest, rangeland and grassland. The depth to $CaCO_3$ serves as a long-term site record of both productivity and precipitation regime. In the Inland Pacific Northwest Region, $CaCO_3$-containing Aridisols support desert plant communities in the driest areas, but are used to produce a variety of crops under irrigation management. Most of the soils in the Snake River Plains contain significant quantities of $CaCO_3$, sometimes leached from upper horizons on older geomorphic surfaces. Soils support mostly sagebrush–grass communities under native conditions. Limited leaching of $CaCO_3$ occurred also in soils of the Lost River Valleys and Mountains and Eastern Idaho Plateaus MLRAs with rangeland as major land use (McDaniel 2017).

The arid and semiarid ecosystems of the LRR Western Range and Irrigated Region contain soils with pedogenic carbonates formed in the past under increased water availability (Rasmussen et al. 2017). The native vegetation consists largely of shrubs, interspersed grasses and scattered trees in the low-lying areas, with areas of forest in the cooler, wetter mountain ranges. Much of the low-lying land in this region is used for grazing with areas of irrigated agricultural production where water is available and soils are suitable. One of the major resource management concerns on croplands includes soil salinity and sodium (Na) content while overgrazing is a concern on areas used for rangeland. In the Basin and Range Province MLRAs, older surfaces tend to contain well-developed soils that exhibit calcic or petrocalcic horizons. The sources of the carbonate reactants include Ca inputs from precipitation, direct $CaCO_3$ influx associated with dust and eolian deposition, and both Ca and HCO_3^- from the chemical weathering of primary minerals in the soil profile. However, dominant source for reactants is a combination of precipitation and dust inputs. For example, up to 90% of the carbonate in a cemented carbonate horizon in the Chihuahuan Desert may be sourced from atmospheric inputs. In general, dust can contribute 0.1–12 g m^{-2} $year^{-1}$ (0.003–0.354 oz yd^{-2} $year^{-1}$) of $CaCO_3$. However, the dominant source of Ca in the Chihuahuan Desert may be precipitation as carbonate precipitation resulting from Ca influx from rain may be two to three times greater than carbonate formed via calcareous dust input. Eolian influx contributes to rapid accumulation of carbonate in Holocene soils of the Mojave Desert. Eolian influx also contributes to the accumulation of carbonates, with dust contributing a large fraction of the Ca for carbonate formation, particularly in the Mojave and Sonoran deserts (Rasmussen et al. 2017).

As the climate gets progressively drier from east to west across the Great Plains, the depth to secondary carbonates decreases because there is less effective moisture to leach carbonates deeper in the soil profile (Hirmas and Mandel 2017). Shortgrass prairie and bunchgrass steppes extend eastward from the foot of the Rocky Mountains in Colorado and Wyoming into Kansas, Nebraska, and South Dakota. These grasslands are dominated by blue grama (*Bouteloua gracilis* (Willd. ex Kunth) Lag. ex Griffiths) and buffalo grass (*Bouteloua dactyloides* (Nutt.) Columbus 1999), but include also larger plants such as yucca (*Yucca* spp.) and prickly pear cactus (*Opuntia* spp.), as well as woody shrubs such as sagebrush. Many soils in the Northern Plains have subsoil accumulations of carbonates (calcic horizons). Cool-season grasses such as fescue (*Festuca* spp.), western wheatgrass (*Pascopyrum smithii* (Rydb.) Á.Löve), and needlegrass (*Stipa* spp.), dominate the prairies. Wetlands are also common in this region, and many have saline soils that support halophytic graminoids such as alkali bulrush (*Bolboschoenus maritimus* (L.) Palla), inland salt grass (*Distichlis spicata* (L.) Greene), and Nuttall's alkali grass (*Puccinellia nuttalliana* (Schult.) Hitchc.), and shrubs such as black greasewood (Sarcobatus vermiculatus (Hook.) Torr.; Hirmas and Mandel 2017).

Common and secondary carbonate may be found in argillic horizons in soil formed in matrix-supported calcareous till on glacial till plains and moraines in north-central South Dakota, central and northwestern North Dakota, and northeastern Montana (Fig. 4.3; Hirmas and Mandel 2017). In the Northern Plains, secondary carbonate accumulation increases and the depth to carbonate decreases in the till-derived soils along the westward transect. Soils formed in glacially derived loess deposits, especially in the northern Central Great Plains, are often calcareous although the depth of the carbonate-free zone at the surface of the profile tends to be a function of rainfall due to leaching of the carbonates. These soils are mostly used for rangeland in the west, and cultivated and irrigated for sorghum and corn toward the east. Soils developed in calcareous alluvium in the Great Plains often exhibit accumulations of $CaCO_3$ along contacts between fine-grained overbank (flood) deposits, and underlying coarse-grained channel and near-channel deposits. Going from east to west across the Great Plains, there is an increase in $CaCO_3$ accumulation in the subsoil of soils formed on similar-age Holocene terraces. The depth of secondary carbonates also decreases from east to west (Hirmas and Mandel 2017).

In the southwestern portion of the Central Great Plains, soils in the south part of the Southern High Plains contain thick calcic horizons (Hirmas and Mandel 2017). Common crops grown are cotton, grain sorghum, and wheat. There are many

Fig. 4.3 Argillic zone alteration from hydrothermal veins (Orphan Boy Mine, Butte, Montana, USA; Creative Commons Attribution 2.0 Generic license)

petrocalcic horizons in the Ogallala Formation of the High Plains. These soils are mostly cultivated, with wheat and grain sorghums as the principal crops. Petrocalcic horizons that are typically 15–90 cm (5.9–35.4 in) thick are also found in the driest areas of the Southern High Plains, including northwestern Texas and eastern New Mexico. These soils are primarily used for rangeland. Soils of Playas are underlain by calcic and petrocalcic horizons. Soils mantling Salina floors contain calcic horizons in the lower subsoil with calcium carbonate equivalents (CCEs) between 40% and 60%. These soils are devoid of vegetation due to the high concentrations of salts and exchangeable Na. The Big Basin in the Eastern Part of the Central Rolling Red Plains, south-central Kansas, has formed shallow, well-drained soils with moderately developed calcic horizons in the underlying lime-cemented sandstone residuum. Soils of Solution Basin walls contain visible secondary carbonates in subsurface horizons (Hirmas and Mandel 2017).

Some of the major soils in MLRAs of the Edwards Plateau and the Texas portion of the Grand Prairie are shallow (<50 cm [<19.7 in]) to petrocalcic horizons (Ruppert 2017). Rainfall decreases from east to west on the Edwards Plateau, and calcic horizons are increasingly found and an increasing strength of expression of calcic horizons. The formation of calcic and petrocalcic horizons on the Edwards plateau may be explained by *in situ* secondary carbonate precipitation directly from the primary carbonates of the parent rock (with no illuviation). Dust inputs or dust retention may be less important for the formation of calcic and petrocalcic horizons. A third model may be more appropriate where petrocalcic horizons formed in skeletal parent

Fig. 4.4 Calcic horizon at about 180 cm (6 ft) depth in a Torrertic Paleustoll (Fine, mixed, superactive, thermic) – Pullman; Potter County, Texas, USA (Creative Commons Attribution 2.0 Generic license)

materials containing limestone clasts or fragments. Such skeletal materials are abundant in the Western and Southern Edwards Plateau and in the Rio Grande Plains. Specifically, illuviation takes place with carbonate fragments forming both the parent material and fabric for petrocalcic horizon formation. Often, the limestone clasts which gave rise to the petrocalcic horizons have been completely dissolved above the petrocalcic leading to non-calcareous soils above the petrocalcic horizon. The typical agricultural land use on the Edwards Plateau is grazing in a landscape covered by trees, shrubs, low shrubs, and short/mid grasses (Ruppert 2017).

Current soil morphologies in the Edwards Plateau may display the preserved or weathered remains of calcic or petrocalcic horizons which were developed in earlier times, with the differential rainfall across the Edwards Plateau playing a role in preserving petrocalcic features in the west and destroying such features in the east (Fig. 4.4; Ruppert 2017). Petrocalcic-bearing soils in the Lampasas Cut Plain of the Grand Prairie are viewed as relict. Some soils of MLRAs in the Rio Grande Plains can be distinguished through the presence of petrocalcic horizons. For example, soils more common to the west of the MLRA Northern Rio Grande Plain commonly exhibit petrocalcic horizons. About 40% of the MLRA Central Rio Grande Plain is underlain by petrocalcic horizons, and for the most part if shallow soils are encountered, it is because of an underlying petrocalcic horizon. Some of the petrocalcic horizons appear to be neo-formed and not relict. The MLRA Lower Rio Grande Plain does not contain an overabundance of petrocalcic horizons. The predominating land uses are row cropping, dryland row cropping and rangeland (Ruppert 2017).

In some soils of the LRR Northern Lake States Forest and Forage Region containing $CaCO_3$ in the parent material, carbonates leach from the upper profile to a meter (39 in) or more (Schaetzl 2017). This process occurs in eastern Upper Peninsula of Michigan, and parts of western Wisconsin and eastern Minnesota. Soils that form in calcareous parent materials tend to be more fertile. Northern hardwood is the dominant vegetation but many areas of agriculture do exist in the region (Schaetzl 2017).

Evaporative discharge caused by upward water movement may contribute to pedogenic carbonate formation in the Rolling Till Prairie, Till Plains, and Loess Uplands MLRAs within the LRR Central Feed Grains and Livestock Region (Libohova et al. 2017). Pedogenic carbonate formation may also occur from $CaCO_3$ transported downslope in soils on glacial outwash and lakebeds. Almost 70% of the area under cropland is used for corn, soybeans, alfalfa, spring wheat, and oats while forested areas occur as narrow strips along streams and rivers or as shelterbelts and windbreaks around farmsteads. Free carbonates generally occurring at a depth of 100–200 cm (39.4–78.7 in) in the Deep Loess and Drift, Iowa and Missouri Heavy Till Plain, and Till Plain MLRAs may contribute to pedogenic carbonate formation. Free carbonates occur generally below a depth of 100 cm (39.4 in) the Thin Loess and till Plains, Central Mississippi Valley Wooded Slopes, Central Claypan Areas, and Cherokee Prairies MLRAs. This is also the case at the Western Michigan Fruit Belt, Southwestern Michigan Fruit and Truck Crop Belt, Southern Michigan and Northern Indiana Drift Plain, Erie-Huron Lake Plain, and Ontario-Erie Plain and Finger Lakes Region MLRAs. The MLRAs underwent a deep transformation with agriculture development that overhauled the prairie vegetation and replaced it mostly with corn, soybean, and other cash crops. Today, the region supports high levels of food and livestock production (Libohova et al. 2017).

Soils in the LRRs East and Central Farming and Forest Region, and Atlantic Basin Diversified Farming Region developed from carbonate bedrock (Lee and Kabrick 2017). Part of the Ca and magnesium (Mg) released during the weathering of the limestone and dolostone was retained in the soil, and potentially contributes to secondary carbonate formation. Land use is forest, grassland and cropland (Lee and Kabrick 2017).

Pedogenic carbonate formation may occur in the LRRs Mississippi Delta Cotton and Feed Grains Region; South Atlantic and Gulf Slope Cash Crops, Forest, and Livestock Region; and Atlantic and Gulf Coast Lowland Forest and Crop Region as some soils in those LRRs are developed from calcareous bedrock (West et al. 2017).

Carbonates commonly occur from 100 to 150 cm (39.4 to 59.1 in) depth in soils of the St. Lawrence-Champlain Plain MLRAs (Wilson and Shaw 2017), and some contribution from pedogenic carbonate formation may occur aside from the calcareous parent materials.

Soils with secondary carbonate deposition have been reported in the warmest and driest parts of northeastern Interior Alaska (Yukon Flats; Ping et al. 2017). Elsewhere reports of pedogenic carbonates in Interior Alaska represent exhumed horizons from buried Pleistocene soils or transient evaporative crusts on river silt bars that do not persist. Soils on high floodplains have carbonates removed from the upper part of the soil (Ping et al. 2017), which may contribute to pedogenic carbonate formation at deeper depths.

4.4 Management of Soil Inorganic Carbon Stocks

The previous sections indicated that SIC derived from lithogenic and/or pedogenic carbonates may occur in terrestrial biomes of the U.S. in arid, semi-arid and other regions. The long global mean turnover time of pedogenic inorganic C of up to 85,000 years indicates that maintaining and/or increasing pedogenic inorganic C stocks is an effective soil C sequestration strategy (Lal and Kimble 2000; Schlesinger 1985). However, some studies have indicated slow formation rates of the SIC stock while others pointed to more dynamic SIC changes (Kim et al. 2020; Berner 2003). For example, SIC can be generated in months to years if the soil environment is suitable (Wang et al. 2016). Soil Ca and pH, in particular, may be managed to purposefully build silicatic pedogenic carbonate to sequester CO_2 (Monger et al. 2015). Some studies on changes in SIC stocks at sites in the U.S., and how soil and land-use management practices including crop cultivation and fertilization may affect SIC will be discussed below (Tables 4.2 and 4.3).

When soils are irrigated with water having high concentrations of Ca^{2+} and/or Mg^{2+}, and those with dissolved HCO_3^- and CO_3^{2-} ions combined with a high pH and an elevated temperature, $CaCO_3$ can precipitate in soils and result in SIC sequestration (Table 4.4). In contrast, the depletion of bicarbonate rich groundwater in the U.S. may contribute to CO_2 emissions into the atmosphere as CO_2 may be released when calcite precipitates in soil after irrigation with bicarbonate rich groundwater (Eq. 4.2; Wood and Hyndman 2017). The soils to 1-m (39.4 in) depth in a temperate semidesert ecosystem in southern Idaho had SIC stocks of up to 156.0 Mg C ha^{-1} (69.6 tn C ac^{-1}) under irrigated cropping systems compared to 95.0 Mg C ha^{-1} (42.4 tn C ac^{-1}) for those under native sagebrush (Entry et al. 2004). Similarly, irrigation water chemistry played an important role in SIC accumulation in cultivated fields in the Imperial Valley in California as irrigation water contained significant amounts of Ca^{2+} and/or CO_3^{2-}, and leaching was not sufficient to remove the reactants. The SIC stocks in the top 1 m (39.4 in) were 208.2 Mg C ha^{-1} (92.9 tn C ac^{-1}) after 85 years of irrigated agriculture and 165.5 Mg C ha^{-1} (73.8 tn C ac^{-1}) under native vegetation (Wu et al. 2008).

When irrigation water chemistry does not support $CaCO_3$ formation and leaching occurs, SIC stocks under irrigated farming may also decrease. For example, SIC stocks in the top 1 m (39.4 in) were 14.9 Mg C ha^{-1} (6.6 tn C ac^{-1}) after 55 years under irrigation agriculture but 25.8 Mg C ha^{-1} (11.5 tn C ac^{-1}) under native

Table 4.2 Soil and land-use management practices for maintaining and/or increasing soil inorganic carbon stocks in terrestrial biomes of the United States

Practice	Reference
Irrigation with water having high concentrations of calcium and/or magnesium ions, and those with dissolved bicarbonate and carbonate ions; managing leaching processes by irrigation management	Wood and Hyndman (2017) and Wu et al. (2008)
Adding calcium and magnesium with dolomite, calcium with gypsum or lime, and/or magnesium with rock minerals	Zamanian et al. (2021) and West and McBride (2005)
Applying organic and mineral amendments resulting in higher soil inputs of calcium, magnesium and/or sodium ions	Lal (2004)
Converting desert shrubs to fertilized and irrigated turfgrass	Kim et al. (2020) and Wang et al. (2016)
Adding demolition waste (particularly cement and concrete) to urban soils	Lorenz and Lal (2015)

Table 4.3 Effects of crop cultivation and fertilization on soil inorganic carbon sequestration in terrestrial biomes of the United States (negative values indicate loss; Kim et al. 2020)

Soil depth		Dominant crop species	Nitrogen fertilization		Soil inorganic carbon sequestration[a]	
cm	in		kg N ha^{-1} $year^{-1}$	lbs N ac^{-1} $year^{-1}$	Mg C ha^{-1} $year^{-1}$	tn C ac^{-1} $year^{-1}$
0–240	0–94	Wheat (*Triticum aestivum* L.)	0	0	−5.38	−2.40
		Corn (*Zea mays* L.), oats (*Avena sativa* L.), wheat, peas (*Pisum sativum* L.)	55	49	−4.49[b]	−2.00[b]
240–490	94–193	Cotton (*Gossypium hirsutum* L.)	34	30	−12.83	−5.72
490–730	94–287	Corn, oats, wheat, peas	55	49	1.11[b]	0.50[b]

[a]Calculated from differences in soil inorganic carbon stocks between soils under natural vegetation and those under rain-fed cultivation
[b]Assuming cultivation for 100 years

vegetation in the San Joaquin Valley California (Wu et al. 2008). Similarly, Sartori et al. (2007) reported a decrease in SIC stocks from 22.2 to 7.3 Mg C ha^{-1} (9.9 tn ac^{-1} to 3.3 tn C ac^{-1}) in 0–50 cm (0–19.7 in) depth with increasing age of fertilized and irrigated poplar (*Populus* spp.) plantations in desert soils of the southern Columbia Plateau of Oregon. This decrease was related to both the inhibition of $CaCO_3$ formation and dissolution along the soil profile. Specifically, decalcification processes may have increased with plantation age because greater amounts of water percolated over time through the desert soils. The older poplar stands had the greatest soil moisture retention and reduced soil water evaporation. Both processes may have reduced the accumulation of $CaCO_3$ under the plantations by enhancing soil moisture retention. Otherwise, deposition of $CaCO_3$ may have also been inhibited by the increase in dissolved organic matter (DOM) and weathering in older stands

Table 4.4 Effects of irrigation practices on soil inorganic carbon sequestration in terrestrial biomes of the United States (negative values indicate loss)

	Soil depth		Soil inorganic carbon sequestration		Reference
Practice	cm	in	Mg C ha^{-1} year^{-1}	tn C ac^{-1} year^{-1}	
Irrigation water chemistry supporting carbonate formation; leaching processes managed	0–100	0–39	0.50[a]	0.22[a]	Wu et al. (2008)
	0–400	0–157	0.12–0.32; 0.95[b]	0.05–0.14; 0.42[b]	Eshel et al. (2007)
Irrigation water chemistry not supporting carbonate formation; leaching not managed	0–50	0–19	−1.86[c]	−0.83[c]	Sartori et al. (2007)
	0–100	0–39	−0.20[a]	−0.09[a]	Wu et al. (2008)
	0–240	0–94	−5.43[a]	−2.42[a]	Kim et al. (2020)
	240–490	94–193	−4.38[d]	−1.96[d]	
			−13.81[a]	−6.16[a]	
	490–730	193–287	−7.30[d]	−3.26[d]	

[a]Difference between irrigated cultivated soils and non-irrigated soils under natural vegetation
[b]Difference between irrigated and non-irrigated cultivated soils; depending on water chemistry and assuming that carbonate dissolution is carbon sequestration
[c]Difference between first- and second-rotation stands
[d]Difference between irrigated and non-irrigated cultivated soils

via dissolution by carbonic and organic acids. Soils under the younger poplar stands may have also been more susceptible to wind erosion/deposition processes that may increase the presence of dust-derived carbonates (Sartori et al. 2007). In addition, N fertilization in combination with soil acidification may have also contributed to the dissolution of $CaCO_3$, and this process may have increased with stand age (Zamanian et al. 2018).

Converting desert shrubs to fertilized and irrigated turfgrass has been shown to increase SIC in desert soils of New Mexiko in less than ten years (Wang et al. 2016). For example, SIC in 10 cm (3.9 in) depth at a desert shrub site had an average value of 1.92 g C kg^{-1} (0.031 oz C lbs^{-1}) while the turfgrass site had an average value of 4.41 g C kg^{-1} (0.071 oz C lbs^{-1}). Thus, SIC may be managed to sequester atmospheric CO_2 in desert biomes in the U.S. However, it also needs to be demonstrated that: (i) the source of Ca is derived directly from silicate minerals (silicatic pedogenic carbonate) rather than preexisting carbonate (calcitic pedogenic carbonate), (ii) the carbonate is durable and not subject to rapid dissolution, and (iii) upstream CO_2 emissions are accounted for (Wang et al. 2016). Further, changes in SIC stocks at deeper soil depths including the fate of any mobilized SIC must also be considered for the assessment of CO_2 sequestration potential by SIC (Table 4.3; Kim et al. 2020).

In addition to contributing to increases in SIC, agricultural land use can also result in SIC losses by enhancing dissolution and leaching of carbonates. For example, less SIC storage under irrigated cropland plots compared to those under native vegetation were reported for deep soil depths at three among five sites in the southern Great Plains in the U.S. (Kim et al. 2020). In 2.4–4.9 m (7.9–16.1 ft) depth, SIC stocks at the sites San Angelo and Goodwell were 1851 and 1103 Mg C ha^{-1} (826 and 492 tn C ac^{-1}) under natural vegetation, and 1299 and 400 Mg C ha^{-1} (579 and 178 tn C ac^{-1}) under irrigated cropland, respectively. The SIC stocks to 2.4-m (7.9-ft) depth at the site Quanah were also lower for irrigated cropland (183 Mg C ha^{-1} [82 tn C ac^{-1}]) compared to those under natural vegetation (509 Mg C ha^{-1} [227 tn C ac^{-1}]). Overall, water rather than differences in land-use management such as tillage or fertilization had a stronger effect on SIC losses. The final fate of the lost SIC could not be assessed with the observations but its understanding is critical in determining whether the SIC mobilization represents a net ecosystem C source or sink. Three possible pathways for the lost SIC are degassing, leaching, or reprecipitation. At the studied sites in the southern Great Plains, reprecipitation of dissolved SIC past the sampling depth (no net C flux), storage of bicarbonate in slow-moving aquifers, or greater interactive effects of agricultural practices on SIC may have occurred. Further evaluation of net sink or source potential of the SIC stock (i.e., fate of dissolved inorganic carbon [DIC]) and associated time lags under changing land uses is, therefore, needed (Kim et al. 2020).

The reaction of soil carbonates with H^+ ions may have led to HCO_3^- and CO_2 formation and, thus, decreases in SIC stocks in 90–120 cm (35–47 in) depth over 11 years at a corn-soybean rotation in Iowa (Dold et al. 2020). Specifically, H^+ ions may have been released in 0–15 cm (0–6 in) by the anhydrous N fertilizer application releasing H+ ions into the soil solution indicated by the observed decrease in soil pH. However, it was unclear whether soil pH also decreased in the subsoil (Dold et al. 2020).

Wen et al. (2021) assessed carbonate weathering using field data and reactive transport numerical experiments at the Konza Prairie Biological Station in Kansas. Woody encroachment into grasslands may deepen root distribution at this prairie site. The study indicated that deepening roots in woodlands that channel more water into un-weathered carbonate at depths may elevate weathering rates by more than an order of magnitude compared to grasslands (Wen et al. 2021).

Calcium supplied as dolomite, gypsum or lime, and Mg from rock minerals are sometimes added to soils, for example, to neutralize soil acidification (Zamanian et al. 2021). About 30 Mt (33 million tn) of lime are annually applied to agricultural soils in the U.S. (West and McBride 2005). The dissolution of agricultural lime ($CaCO_3$) and hydrated lime can be either a net source or sink of CO_2 (Kunhikrishnan et al. 2016). For example, lime-derived CO_2 reacts with microbial respiration derived H_2CO_3 to produce carbonate material, which is a sink for CO_2 in soil. Agricultural lime is either a net sink for CO_2 in calcareous soils with high pH, but it is a net source of CO_2 in acidic soils. Adding lime to soils can generally contribute to C sequestration by increasing methane (CH_4) oxidation and reducing greenhouse gas (GHG) emissions (Kunhikrishnan et al. 2016). For example, liming acidic soils to neutrality may decrease soil nitrous oxide (N_2O) emissions (Hénault et al. 2019). Thus, depending on whether the reaction of lime in soil is with strong acids or with carbonic acid, agricultural lime can either be a source or a sink for CO_2 (Hamilton et al. 2007). Liming during acidity neutralization may cause CO_2 losses of 273 Tg C year^{-1} (301 million tn C year^{-1}; Raza et al. 2021). A recent global meta-analysis indicated that the overall GHG balance of liming acidic soils is likely neutral (Wang et al. 2021). Specifically, the CO_2 release (624–656 Tg CO_2 year^{-1} [688–723 million tn CO_2 year^{-1}]) derived from lime mining, transport and application, and lime dissolution may offset reductions in soil N_2O emissions (0.60–0.67 Tg N_2O-N year^{-1} [0.66–0.74 million tn N_2O-N year^{-1}]), paddy soil CH_4 emissions (1.75–2.21 Tg CH_4 year^{-1} [1.93–2.44 million tn CH_4 year^{-1}]), and increases in SOC stocks (65.7–110 Tg C year^{-1} [72–121 million tn C year^{-1}]; Wang et al. 2021).

Application of organic and mineral amendments resulting in higher soil inputs of Ca^{2+}, Mg^{2+} and/or Na^+ may enhance the formation of pedogenic carbonates (Lal 2004). These cations are sometimes added with livestock manure (Rayne and Aula 2020). Adding cations may contribute to soil C sequestration when these amendments are supplied from outside the ecosystem via fertilizers or other soil amendments (Nordt et al. 2000). For example, the soil application of rock dust containing crushed silicate rock may sequester CO_2 as bicarbonate and carbonate minerals (abiotic CO_2 removal [CDR] pathway; Kantola et al. 2017). About 0.5–1.0 Mg CO_2 (0.6–1.1 tn CO_2) may be sequestered per one Mg (1.1 tn) of silicate rock applied by this process of enhanced weathering (Moosdorf et al. 2014). However, basalt also improves soil fertility and potentially enhances ecosystem C storage (biotic CDR pathway; Goll et al. 2021). Including this previously overlooked enhancement of biological C sequestration strongly increases the global CO_2 removal potential of basalt application to soil. Thus, to sustain a global CDR of 1.3 Pg CO_2 year^{-1} (1.4 billion tn CO_2 year^{-1}), a minimum amount of 1.1 Pg year^{-1} (1.2 billion tn year^{-1}) of basalt may be sufficient if applied to regions with the largest CDR potential (Goll et al. 2021). Instead of basalt, volcanic ash or tephra is also discussed as a soil amendment for accumulating both SIC and SOC (Minasny et al. 2021).

The SIC stocks of urban ecosystems in the U.S. may be enhanced by soil and land-use management practices (Lorenz and Lal 2015). Adding demolition waste (particularly cement and concrete), for example, may contribute to urban SIC stocks. Options to enhance SIC stocks under existing urban vegetation cover include optimizing productivity by fertilization (i.e., applying NH_4^+-producing fertilizers) and irrigation (i.e., with alkaline water). Establishing a new vegetation cover on bare urban soil after construction activities is also an opportunity for enhancing SIC stocks. For example, either planting should be done in fertile urban soil, or soil fertility should be improved during site establishment by adding Ca- and Mg-rich OM, inorganic fertilizers containing Ca and Mg, or waste materials containing Ca- and Mg-silicate minerals. Plant species with an extensive surficial root system producing high amounts of Ca- and Mg-rich litter should be established. Before planting, soil structure should be improved by deep tillage. However, soil disturbance should be reduced after plant establishment to allow the development of well-structured soil, and the proliferation of deep-burrowing earthworms. While surface soil can be acidified by applying NH_4^+-producing fertilizers, burial of alkaline compounds at deeper depths assists in creating alkaline conditions for carbonate precipitation in urban subsoils (Lorenz and Lal 2015). Indirect effects of the urban environment on SIC sequestration may be the deposition of Ca- or Mg-containing dust which can contribute to pedogenic carbonate formation. Further, magnesium chloride ($MgCl_2$) and calcium chloride ($CaCl_2$) are applied to roads in colder urban areas in the U.S. to remove snow and prevent ice buildup. Some residues of de-icing salts are retained in urban soils (Shannon et al. 2020), and this may also contribute to pedogenic carbonate formation.

4.5 Conclusions

Lithogenic carbonates contribute to the SIC stock in terrestrial biomes of the U.S., and they are widely distributed in the form of calcareous bedrock. The other principal SIC form are pedogenic carbonates which may also occur widely in U.S. soils. However, only the formation of silicatic pedogenic carbonates contributes to soil C sequestration when the Ca^{2+} precipitated in carbonates is released directly from silicates. In contrast to SOC sequestration, this process can proceed over hundreds to thousands of years and is, thus, an effective climate change mitigation mechanism. The SIC stock in the form of carbonates influences also soil microbial and enzyme activities, the rates of SOC mineralization, soil pH, adsorption-desorption processes, soil cementation, acidity buffering, aggregate formation and stabilization, nutrient cycling and availability, and water permeability and plant productivity. Soil and land-use management practices affect SIC dynamics but how those practices can be targeted to reduce SIC losses and increase SIC formation is only recently receiving more attention. However, soil and land-use management practices to enhance soil C sequestration must also include SIC sequestration due to the importance of the SIC stock for climate change adaptation and mitigation, and other environmental benefits.

References

An H, Wu X, Zhang Y, Tang Z (2019) Effects of land-use change on soil inorganic carbon: a meta-analysis. Geoderma 353:273–282. https://doi.org/10.1016/j.geoderma.2019.07.008

Bar-On YM, Phillips R, Milo R (2018) The biomass distribution on Earth. P Natl Acad Sci USA 115:6506–6511. www.pnas.org/cgi/doi/10.1073/pnas.1711842115

Berner RA (2003) The long-term carbon cycle, fossil fuels and atmospheric composition. Nature 426:323–326. https://doi.org/10.1038/nature02131

Bockheim JG (2017) Soils of the USA: the broad perspective. In: West LT, Singer MJ, Hartemink AE (eds) The soils of the USA, World Soils Book Series. Springer, Cham, pp 43–75. https://doi.org/10.1007/978-3-319-41870-4_4

Collins ME (2017) Soils of Florida and the Caribbean: LRRs U and Z. In: West LT, Singer MJ, Hartemink AE (eds) The soils of the USA, World Soils Book Series. Springer, Cham, pp 281–303. https://doi.org/10.1007/978-3-319-41870-4_15

Ditzler CA (2017) Soil properties and classification (soil taxonomy). In: West LT, Singer MJ, Hartemink AE (eds) The soils of the USA, World Soils Book Series. Springer, Cham, pp 29–41. https://doi.org/10.1007/978-3-319-41870-4_3

Dold C, Wacha KM, Sauer TJ, Hatfield JL, Prueger JH (2020) Measured and simulated carbon dynamics in Midwestern U.S. corn-soybean rotations. Global Biogeochem Cy 35:e2020GB006685. https://doi.org/10.1029/2020GB006685

Dürr HH, Meybeck M, Dürr SH (2005) Lithologic composition of the Earth's continental surfaces derived from a new digital map emphasizing riverine material transfer. Global Biogeochem Cycles 19:GB4S10. https://doi.org/10.1029/2005GB002515

Entry JA, Sojka RE, Shewmaker GE (2004) Irrigation increases inorganic carbon in agricultural soils. Environ Manage 33:S309–S317. https://doi.org/10.1007/s00267-003-9140-3

Eshel G, Fine P, Singer MJ (2007) Total soil carbon and water quality: an implication for carbon sequestration. Soil Sci Soc Am J 71:397–405. https://doi.org/10.2136/sssaj2006.0061

Eswaran H, Reich PF, Kimble JM et al (2000) Global carbon stocks. In: Lal R, Kimble JM, Eswaran H, Stewart BA (eds) Global climate change and pedogenic carbonates. CRC Press, Boca Raton, pp 15–25

Garcia C, Nannipieri P, Hernandez T (2018) The future of soil carbon. In: Garcia C, Nannipieri P, Hernandez T (eds) The future of soil carbon – its conservation and formation. Academic Press, Oxford, pp 239–267. https://doi.org/10.1016/B978-0-12-811687-6.00009-2

Goll DS, Ciais P, Amann T et al (2021) Potential CO_2 removal from enhanced weathering by ecosystem responses to powdered rock. Nat Geosci 14:545–549. https://doi.org/10.1038/s41561-021-00798-x

Graham RC, Indorante SJ (2017) Concepts of soil formation and soil survey. In: West LT, Singer MJ, Hartemink AE (eds) The soils of the USA, World Soils Book Series. Springer, Cham, pp 9–27. https://doi.org/10.1007/978-3-319-41870-4_2

Guo Y, Amundson R, Gong P, Yu Q (2006a) Quantity and spatial variability of soil carbon in the conterminous United States. Soil Sci Soc Am J 70:590–600. https://doi.org/10.2136/sssaj2005.0162

Guo Y, Gong P, Amundson R, Yu Q (2006b) Analysis of factors controlling soil carbon in the conterminous United States. Soil Sci Soc Am J 70:601–612. https://doi.org/10.2136/sssaj2005.0163

Hamilton SK, Kurzman AL, Arango C, Jin L, Robertson GP (2007) Evidence for carbon sequestration by agricultural liming. Global Biogeochem Cy 21:GB2021. https://doi.org/10.1029/2006GB002738

Hénault C, Bourennane H, Ayzac A et al (2019) Management of soil pH promotes nitrous oxide reduction and thus mitigates soil emissions of this greenhouse gas. Sci Rep 9:20182. https://doi.org/10.1038/s41598-019-56694-3

Hirmas DR, Mandel RD (2017) Soils of the Great Plains. In: West LT, Singer MJ, Hartemink AE (eds) The soils of the USA, World Soils Book Series. Springer, Cham, pp 131–163. https://doi.org/10.1007/978-3-319-41870-4_8

Kantola IB, Masters MD, Beerling DJ, Long SP, DeLucia EH (2017) Potential of global croplands and bioenergy crops for climate change mitigation through deployment for enhanced weathering. Biol Lett 13:20160714. https://doi.org/10.1098/rsbl.2016.0714

Kim JH, Jobbágy EG, Richter DD, Trumbore SE, Jackson RB (2020) Agricultural acceleration of soil carbonate weathering. Glob Change Biol 26:5988–6002. https://doi.org/10.1111/gcb.15207

Kleber M, Bourg IC, Coward EK et al (2021) Dynamic interactions at the mineral–organic matter interface. Nat Rev Earth Environ 2:402–421. https://doi.org/10.1038/s43017-021-00162-y

Kunhikrishnan A, Thangarajan R, Bolan NS et al (2016) Functional relationships of soil acidification, liming, and greenhouse gas flux. Adv Agron 139:1–71. https://doi.org/10.1016/bs.agron.2016.05.001

Lal R (2004) Carbon sequestration in dryland ecosystems. Environ Manage 33:528–544. https://doi.org/10.1007/s00267-003-9110-9

Lal R, Kimble JM (2000) Pedogenic carbonates and the global carbon cycle. In: Lal R, Kimble JM, Eswaran H, Stewart BA (eds) Global climate change and pedogenic carbonates. CRC Press, Boca Raton, pp 291–302

Lee BD, Kabrick JM (2017) East and Central Farming and Forest Region and Atlantic Basin Diversified Farming Region: LRRs N and S. In: West LT, Singer MJ, Hartemink AE (eds) The soils of the USA, World Soils Book Series. Springer, Cham, pp 227–241. https://doi.org/10.1007/978-3-319-41870-4_12

Libohova Z, Schoeneberger P, Wysocki D, Owens PR (2017) Soils of the Central Feed Grains and Livestock Region and Lake State Fruit, Truck Crop, and Dairy Region USA: LRRs M and L. In: West LT, Singer MJ, Hartemink AE (eds) The soils of the USA, World Soils Book Series. Springer, Cham, pp 209–226. https://doi.org/10.1007/978-3-319-41870-4_11

Liu Z, Sun Y, Zhang Y et al (2021) Soil microbes transform inorganic carbon into organic carbon by dark fixation pathways in desert soil. J Geophys Res Biogeo 126:e2020JG006047. https://doi.org/10.1029/2020JG006047

Lorenz K, Lal R (2015) Managing soil carbon stocks to enhance the resilience of urban ecosystems. Carbon Manage 6:35–50. https://doi.org/10.1080/17583004.2015.1071182

McDaniel PA (2017) Soils of the Rocky Mountain and Inland Pacific Northwest Region: LRRs B and E. In: West LT, Singer MJ, Hartemink AE (eds) The soils of the USA, World Soils Book Series. Springer, Cham, pp 101–114. https://doi.org/10.1007/978-3-319-41870-4_6

Minasny B, Fiantis D, Hairiah K, Van Noordwijk M (2021) Applying volcanic ash to croplands – the untapped natural solution. Soil Secur 3:100006. https://doi.org/10.1016/j.soisec.2021.100006

Monger HC (2014) Soils as generators and sinks of inorganic carbon in geologic time. In: Hartemink AE, McSweeney K (eds) Soil carbon. Progress in soil science. Springer, Cham, pp 27–36. https://doi.org/10.1007/978-3-319-04084-4_3

Monger HC, Kraimer RA, Khresat S et al (2015) Sequestration of inorganic carbon in soil and groundwater. Geology 43:375–378. https://doi.org/10.1130/G36449.1

Moosdorf N, Renforth P, Hartmann J (2014) Carbon dioxide efficiency of terrestrial enhanced weathering. Environ Sci Technol 48:4809–4816. https://doi.org/10.1021/es4052022

Nordt LC, Wilding LP, Drees LR (2000) Pedogenic carbonate transformations in leaching soil systems: implications for the global C cycle. In: Lal R, Kimble JM, Eswaran H, Stewart BA (eds) Global climate change and pedogenic carbonates. CRC Press, Boca Raton, pp 43–64

Oades JM, Waters AG (1991) Aggregate hierarchy in soils. Austr J Soil Res 29:815–828

Ping CL, Clark MH, D'Amore D, Michaelson GJ, Swanson DK (2017) In: West LT, Singer MJ, Hartemink AE (eds) The soils of the USA, World Soils Book Series. Springer, Cham, pp 329–350. https://doi.org/10.1007/978-3-319-41870-4_17

Rasmussen C, Lybrand RA, Orem C, Kielhofer J, Holleran M (2017) Soils of the Western range and Irrigated land resource region: LRR D. In: West LT, Singer MJ, Hartemink AE (eds) The soils of the USA, World Soils Book Series. Springer, Cham, pp 115–130. https://doi.org/10.1007/978-3-319-41870-4_7

Rayne N, Aula L (2020) Livestock manure and the impacts on soil health: a review. Soil Syst 4:64. https://doi.org/10.3390/soilsystems4040064

Raza S, Miao N, Wang P et al (2020) Dramatic loss of inorganic carbon by nitrogen-induced soil acidification in Chinese croplands. Glob Change Biol 26:3738–3751. https://doi.org/10.1111/gcb.15101

Raza S, Zamanian K, Ullah S et al (2021) Inorganic carbon losses by soil acidification jeopardize global efforts on carbon sequestration and climate change mitigation. J Clean Prod 315:128036. https://doi.org/10.1016/j.jclepro.2021.128036

Robotham MP, Gavenda RT, Smith CW (2017) Soils of Hawaii and the US Territories in the Pacific: LRRs Q and V. In: West LT, Singer MJ, Hartemink AE (eds) The soils of the USA, World Soils Book Series. Springer, Cham, pp 305–327. https://doi.org/10.1007/978-3-319-41870-4_16

Rowley MC, Grand S, Spangenberg JE, Verrecchia EP (2021) Evidence linking calcium to increased organo-mineral association in soils. Biogeochemistry 153:223–241. https://doi.org/10.1007/s10533-021-00779-7

Ruppert DE (2017) Southwest plateaus and plains range and Cotton region and Southwestern Prairies Cotton and Forage region: LRRs I and J. In: West LT, Singer MJ, Hartemink AE (eds) The soils of the USA, World Soils Book Series. Springer, Cham, pp 165–190. https://doi.org/10.1007/978-3-319-41870-4_9

Sartori F, Lal R, Ebinger MH, Eaton JA (2007) Changes in soil carbon and nutrient pools along a chronosequence of poplar plantations in the Columbia Plateau, Oregon, USA. Agric Ecosyst Environ 122:325–339. https://doi.org/10.1016/j.agee.2007.01.026

Schaetzl RJ (2017) Soils of the Northern Lake States Forest and Forage Region: LRR K. In: West LT, Singer MJ, Hartemink AE (eds) The soils of the USA, World Soils Book Series. Springer, Cham, pp 191–208. https://doi.org/10.1007/978-3-319-41870-4_10

Schlesinger WH (1985) The formation of caliche in soils of the Mojave Desert, California. Geochim Cosmochim Ac 49:57–66. https://doi.org/10.1016/0016-7037(85)90191-7

Schlesinger WH (2000) Carbon sequestration in soils: some cautions amidst optimism. Agric Ecosyst Environ 82:121–127. https://doi.org/10.1016/S0167-8809(00)00221-8

Shannon TP, Ahler SJ, Mathers A, Ziter CD, Dugan HA (2020) Road salt impact on soil electrical conductivity across an urban landscape. J Urban Ecol:1–8. https://doi.org/10.1093/jue/juaa006

Song XD, Yang F, Wu HY et al (2021) Significant loss of soil inorganic carbon at the continental scale. Nat Sci Rev nwab:120. https://doi.org/10.1093/nsr/nwab120

Southard SB, Southard RJ, Burns S (2017) Soils of the Pacific Coast Region: LRRs A and C. In: West LT, Singer MJ, Hartemink AE (eds) The soils of the USA, World Soils Book Series. Springer, Cham, pp 77–100. https://doi.org/10.1007/978-3-319-41870-4_5

Totsche KU, Amelung W, Gerzabek MH et al (2018) Microaggregates in soil. J Plant Nutr Soil Sci 181:104–136. https://doi.org/10.1002/jpln.201600451

Wang JP, Monger C, Wang XJ, Serena M, Leinauer B (2016) Carbon sequestration in response to grassland–shrubland–turfgrass conversions and a test for carbonate biomineralization in desert soils, New Mexico, USA. Soil Sci Soc Am J 80:1591–1603. https://doi.org/10.2136/sssaj2016.03.0061

Wang Y, Yao Z, Zhan Y et al (2021) Potential benefits of liming to acid soils on climate change mitigation and food security. Glob Change Biol 27:2807–2821. https://doi.org/10.1111/gcb.15607

Wen H, Sullivan PL, Macpherson GL, Billings SA, Li L (2021) Deepening roots can enhance carbonate weathering by amplifying CO_2-rich recharge. Biogeosciences 18:55–75. https://doi.org/10.5194/bg-18-55-2021

West TO, McBride AC (2005) The contribution of agricultural lime to carbon dioxide emissions in the United States: dissolution, transport, and net emissions. Agric Ecosyst Environ 108:145–154

West LT, Shaw JN, Mersiovsky EP (2017) Soils of the Southeastern USA: LRRs O, P, and T. In: West LT, Singer MJ, Hartemink AE (eds) The soils of the USA, World Soils Book Series. Springer, Cham, pp 243–260. https://doi.org/10.1007/978-3-319-41870-4_13

Wilson MA, Shaw RK (2017) Northeastern Forage and forest region: LRR R. In: West LT, Singer MJ, Hartemink AE (eds) The soils of the USA, World Soils Book Series. Springer, Cham, pp 261–279. https://doi.org/10.1007/978-3-319-41870-4_14

Wood WW, Hyndman DW (2017) Groundwater depletion: A significant unreported source of atmospheric carbon dioxide. Earth's Future 5:1133–1135. https://doi.org/10.1002/2017EF000586

Wu L, Wood Y, Jiang P et al (2008) Carbon sequestration and dynamics of two irrigated agricultural soils in California. Soil Sci Soc Am J 72:808–814. https://doi.org/10.2136/sssaj2007.0074

Zamanian K, Kuzyakov Y (2019) Contribution of soil inorganic carbon to atmospheric CO_2: more important than previously thought. Glob Change Biol 25:e1–e3. https://doi.org/10.1111/gcb.14463

Zamanian K, Pustovoytov K, Kuzyakov Y (2016) Pedogenic carbonates: forms and formation processes. Earth-Sci Rev 157:1–17. https://doi.org/10.1016/j.earscirev.2016.03.003

Zamanian K, Zarebanadkouki M, Kuzyakov Y (2018) Nitrogen fertilization raises CO_2 efflux from inorganic carbon: a global assessment. Glob Change Biol 24:2810–2817. https://doi.org/10.1111/gcb.14148

Zamanian K, Zhou J, Kuzyakov Y (2021) Soil carbonates: the unaccounted, irrecoverable carbon source. Geoderma 384:114817. https://doi.org/10.1016/j.geoderma.2020.114817

Chapter 5
Incentivizing Soil Organic Carbon Management in Terrestrial Biomes of the United States of America

Abstract The soil organic carbon (SOC) stock of actively managed portions of terrestrial biomes of the U.S. can be maintained and/or increased by targeted soil and land-use management practices. Soils that have lost large amount of SOC in the past (carbon 'debt') such as cropland soils, and soils with unsaturated reactive mineral/metal-associated C pools can be particularly managed towards increasing SOC stocks. This increase is associated with many benefits including improvements in soil health and quality, soil water storage, biodiversity and biomass production while also contributing to climate change adaptation and mitigation. However, the adoption of SOC-conserving practices such as no-till (NT) and cover crops in the U.S. has been slow as these are also associated with costs to farmers, growers, ranchers and forest land owners while their economic benefits are not immediately tangible, in addition to other barriers to adoption. Here, federal and state level legislation can support the more widespread adoption of carbon farming practices in agricultural and forest land. Another lever is the increasing private sector interest in soil health and climate-friendly supply chains which may be facilitated by generating carbon offsets or credits by enhancing SOC stocks. Many companies in the U.S. have realized this potential as a business opportunity but voluntary carbon markets are in a nascent state and outcomes with regards to net zero emission commitments by corporations uncertain. Importantly, SOC credits must be standardized to build trust and confidence in the credits generated. Another approach may be adding price premiums on products from climate-smart and SOC-friendly practices indicated by a unique seal. This would have the advantage of being independent of any political change similar to the USDA seal for products from organic agriculture (OA) which is well respected in society and already in place for several decades. Consumers are becoming increasingly aware about the effects of their personal choices on the environment, and this should be leveraged for incentivizing more climate-smart and SOC-friendly consumption and use of resources. Long-term commitments over decades and longer are needed to increase and stabilize SOC stocks of terrestrial biomes in the U.S. towards adaption and mitigation of climate change. This chapter briefly summarizes evidence from the previous chapters on how the SOC stock of terrestrial biomes in the U.S. is affected by climate and global changes, and how it can be protected and enhanced by soil and land-use manage-

K. Lorenz, R. Lal, *Soil Organic Carbon Sequestration in Terrestrial Biomes of the United States*, https://doi.org/10.1007/978-3-030-95193-1_5

ment practices to increase SOC sequestration. This is followed by an overview on initiatives at the federal and state level, and in industry, private and public sectors to increase the adoption of SOC-enhancing practices. The chapter concludes with a discussion on the potential of carbon farming and carbon markets to incentivize stewardship of SOC by land managers.

Keywords Climate change · Land use · Land cover · Soil organic carbon stock · Soil and land-use management practices · Federal and state policies · Industry, public and private sector initiatives · Carbon farming · Carbon markets · Carbon pricing

5.1 Introduction

The terrestrial biomes of the United States of America (U.S.; 50 states, District of Columbia, Caribbean Islands, and Pacific Islands) contain multiple ecosystems that have common physical properties such as climate and geology. Grassland, pasture and rangeland ecosystems cover an estimated 2.7 million km^2 (1.0 million mi^2), forest ecosystems 2.6 million km^2 (1.0 million mi^2), cropland ecosystems 1.6 million km^2 (0.6 million mi^2), and the total developed area including urban ecosystems in the conterminous U.S. (CONUS, i.e., the 48 adjoining U.S. states on the continent of North America) is estimated at 430,000 km^2 (165,000 mi^2; Bigelow and Borchers 2017; Homer et al. 2020).

This chapter briefly summarizes evidence from the previous chapters on how the SOC stock of terrestrial biomes in the U.S. is affected by climate and global changes, and how it can be protected and enhanced by soil and land-use management practices to increase SOC sequestration. This is followed by an overview on initiatives at the federal and state level, and in industry, private and public sec-tors to increase the adoption of SOC-enhancing practices. The chapter concludes with a discussion on the potential of carbon farming and carbon markets to incentivize stewardship of SOC by land managers.

5.1.1 Climate Change

The climate of the U.S. is changing rapidly (Wuebbles et al. 2017), and this may profoundly affect ecosystems and terrestrial biomes. Annual average temperature over CONUS has increased by 1.0 °C (1.8 °F) in the period 1901–2016. In 2021–2050, annual average temperatures are expected to rise by about 1.4 °C (2.5 °F), relative to the average from 1976 to 2005, under all plausible future climate scenarios. It is extremely likely that human activities, especially emissions of greenhouse gases (GHGs), are the dominant cause of the observed warming since the mid-twentieth century. Changes in the characteristics of extreme events are particularly important for natural ecosystems and biomes. For example, heavy rainfall is increasing in intensity and frequency across the U.S., and is expected to continue to increase. The largest observed changes have occurred in the Northeast. Heatwaves have become more frequent since the 1960s, while extreme cold temperatures and cold waves are less frequent. Recent record-setting hot years are projected to become common in the near future, as annual average temperatures continue to rise. The incidence of large forest fires in the western U.S. and Alaska has increased since the early 1980s, and is projected to further increase in those regions as the climate changes, with profound changes to regional ecosystems. Annual trends toward earlier spring melt and reduced snowpack are already affecting water resources in the western U.S., and these trends are expected to continue. Under higher emission scenarios, and assuming no change to current water resources management, chronic, long-duration hydrological drought is increasingly possible before the end of the twenty-first century (Wuebbles et al. 2017).

The climate of the U.S. is strongly connected to the changing global climate (Wuebbels et al. 2017). The magnitude of climate change beyond the next few decades will depend primarily on the amount of GHGs (especially carbon dioxide [CO_2]) emitted globally. Without major reductions in GHG emissions, the increase in annual average global temperature relative to preindustrial times could reach 5 °C (9 °F) or more by the end of the twenty-first century. With significant reductions in GHG emissions, the increase in annual average global temperature could be limited to 2 °C (3.6 °F) or less. The global atmospheric CO_2 concentration has now passed 400 parts per million (ppm), a level that last occurred about 3 million years ago. Continued growth in CO_2 emissions over the twenty-first century and beyond would lead to an atmospheric CO_2 concentration not experienced in tens to hundreds of millions of years. There is broad consensus that the further and the faster the Earth system is pushed towards warming, the greater the risk of unanticipated changes and impacts, some of which are potentially large and irreversible. The observed increase in carbon (C) emissions over the past 15–20 years has been consistent with higher emissions pathways. In 2014 and 2015, emission growth rates slowed as economic growth became less C-intensive. However, even if this slowing trend continues, it is not yet at a rate that would limit global average temperature change to well below 2 °C (3.6 °F) above preindustrial levels (Wuebbles et al. 2017).

5.1.2 Climate-Induced Changes in Productivity and Soil Organic Carbon

Plant productivity in the U.S. may have not increased with the increased number of frost-free days or with the longer growing season due to plant-specific temperature thresholds, plant–pollinator dependence, and seasonal limitations in water and nutrient availability (Hibbard et al. 2017). Thus, plant-derived soil C inputs may also not have been affected by climate change. Similarly, the net effects of climate change on SOC stocks in the U.S. are uncertain. Higher soil temperatures may promote decomposition and accelerate SOC losses when soil moisture is not limiting microbial activity (Chen et al. 2020). Otherwise, increased soil-moisture variability may eventually reduce the land C sink in the U.S., and result in decreased SOC stocks, particularly, in the western and south-eastern U.S. (Green et al. 2019).

Land-use and land-cover changes produce physical changes in land surface albedo, latent and sensible heat, and atmospheric aerosol and GHG concentrations (Hibbard et al. 2017). These land-use and land-cover changes have turned the terrestrial biosphere (soil and plants) into a net "sink" for C globally in recent decades, and this sink has steadily increased since 1980 (Wuebbels et al. 2017). The continental C sink of the U.S. for 2004–2013 has been estimated at 606–699 Tg C y^{-1} (668–771 million tn C y^{-1}) including terrestrial biomes, inland waters, tidal wetlands and estuaries (Hayes et al. 2018). Forest biomes are important C sinks with more than half of the forest C stock found in soils (U.S. EPA 2021).

Improving soil health involves enhancing SOC (Toor et al. 2021). Specifically, healthy soils can sequester significant amounts of atmospheric CO_2 primarily through increases, and protection and stabilization of the SOC stock. Thus, healthy soils with site-specific SOC stocks contribute to mitigating CO_2-related climate change (Lal 2016). Historically, terrestrial biomes and native ecosystems in the U.S. had higher SOC stocks than the same regions store today. Losses occurred by anthropogenic conversion to managed land, particularly cropland, primarily over the past 200–300 years as human population and land transformation increased substantially (National Academies of Sciences, Engineering, and Medicine 2019). The estimated total SOC losses for the U.S. are 12.2 Pg C (13.4 billion tn C) in 0–200 cm (0–79 in) depth, respectively (Sanderman et al. 2017). Nolan et al. (2021) used the silo conceptualizations of terrestrial C storage centered on refilling past C losses from land use change and land management. Specifically, these estimates of historical losses indicate a hypothetical though highly impractical upper bound for restoring SOC stocks, and their particulate and mineral-associated organic matter (POM and MAOM) fractions through adjustments to land management, specifically, in agroecosystems (Luo et al. 2020). Further, many forest and grassland ecosystems in the U.S. may have additional capacity for protection of SOC by minerals or metals as their reactive mineral/metal-associated C pools are not saturated (Yu et al. 2021). These soils with non-saturated MAOM fractions should be managed for SOC sequestration.

The SOC stock of terrestrial biomes in the U.S. is affected by climate and global changes but can also be protected and enhanced by management. In manageable portions of terrestrial biomes, targeted soil and land-use management practices can be implemented to maintain and enhance SOC stocks (Tables 3.3 and 5.1). Some of the considered options in different terrestrial biomes are summarized below. The Chapter concludes with a discussion of industry, public and private sector initiatives, and federal and state U.S. policies with potential effects on maintaining and/or increasing SOC stocks for climate change adaptation and mitigation, and for improving the environment.

Table 5.1 'Hotspots' of soil organic carbon sequestration in the United States

Terrestrial biome/ecosystem/land use	'Hotspot'	Area		References
		Million km^2	Million mi^2	
Cropland	Highly erodible land used for crop production	0.26	0.10	Sperow (2020)
	Corn Belt of the midwestern United States without remaining A-horizon	0.14	0.05	Thaler et al. (2021)
	"Available" cropland soils neither high in SOC nor sandy	No data	No data	Zomer et al. (2017)
	Croplands with low/now perennial and/or tree cover	No data	No data	Hristov et al. (2018) and Schoeneberger et al. (2017)
Degraded lands, croplands, grasslands	Manageable area	0.73–1.42	0.28–0.55	Chambers et al. (2016)
Forest	Soils with oxalate-extractable aluminum >0.5%	No data	No data	Yu et al. (2021)
Grassland/shrubland	Soils with ammonium acetate exchangeable calcium plus exchangeable magnesium >30 cmol kg^{-1}	No data	No data	
Marginal land	Manageable area	0.55–1.73	0.21–0.67	Yang et al. (2020)
Peatlands/wetlands	Drained area	No data	No data	Kolka et al. (2018)

5.2 Forest Biomes

Nearly all forests in CONUS are considered managed lands (Domke et al. 2018), while in Alaska less accessible and less productive areas in the boreal forest/taiga biome are not directly affected by forest management (Schuur et al. 2018). In total, an estimated 2.8 million km^2 (1.1 million mi^2) of forest land in CONUS and Alaska are considered managed (U.S. EPA 2021). Management is mainly for wood products, water, and recreation services, with C uptake as a secondary outcome (Domke et al. 2018). In many regions, forest C stocks are recovering from historical clearing and thinning dating back to as early as the 1600s. This recovery stimulates C uptake from both afforestation and C accumulation in still-maturing stands. Forest management also has generally accelerated C accumulation rates but the net effects of management activities on forest C stocks and fluxes are unclear. Fire suppression activities have tended to increase forest C stocks. Fuel reduction treatments (e.g., prescribed fire and thinning) may contribute to short-term C losses but collectively to greater long-term C storage than that of untreated forest stands (Domke et al. 2018). Management of forest C in the U.S. must also be adapted to the uncertain effects of climate change (Park et al. 2014).

The forest SOC stock can be maintained when soil C inputs (i.e., NPP or net primary production) and losses are balanced, and stand-replacing disturbances are managed. A general overview of forest management practices potentially contributing to SOC sequestration and increases in SOC stocks is provided in Chap. 3. In 2019, forest land, harvested wood products, woodlands, and urban trees in settlements collectively represented an estimated net uptake of 0.78 Pg CO_2e y^{-1} (0.86 billion tn CO_2e y^{-1}; U.S. EPA 2021). Forest land remaining forest land in CONUS and Alaska, in particular, is a net C sink with an estimated net uptake of 0.58 Pg CO_2e y^{-1} (0.64 billion tn CO_2e y^{-1}) in 2019 (U.S. EPA 2021). Some of this C uptake may contribute to maintaining and/or increasing SOC stocks. For the least accessible of the forest biomes, i.e., the boreal forest/taiga biome, management effects on SOC stocks are poorly understood. It is, however, generally thought that management options to increase boreal forest SOC stocks are limited. For maintaining boreal forest SOC stocks, Nave et al. (2019) summarized the necessity to: (i) decrease fire extent/severity by incorporating young stands into landscape mosaic; and (ii) lighter harvest removals (thin *vs.* clearcut, retain residues and stumps). Management options to increase boreal forest SOC stocks include: (i) fell residual wood to drive paludification; (ii) transition from spruce (*Picea* spp.) to fast growing aspen (*Populus* spp.) with high SOC or fire adapted pines (*Pinus* spp.) with fast SOC recovery; (iii) fertilization to improve tree growth/litter inputs and inhibition of decomposition to increase O horizons; (iv) encouraging forest establishment on mountain tundra; and (v) reforest burned slopes (Nave et al. 2019).

More accessible and productive than the boreal forest/taiga biome is the temperate coniferous forest biome. Thus, nearly all of its area in CONUS is considered managed land (Domke et al. 2018). The moist climate and warm temperate regions in this biome benefits increases in SOC stocks (Sperow 2020). Specifically, the high net ecosystem carbon balance (NECB) of temperate coniferous forests may be conducive for maintaining and/or increasing SOC stocks when key SOC vulnerabilities such as harvest and wildfire are managed. Management practices to maintain SOC stocks in the temperate coniferous forest biome include: (i) fuel management including the use of prescribed fire, (ii) reforestation of burned or harvested forests, (iii) selection of high-SOC species if native forest is converted, (iv) retaining harvest residues, and (v) compensating biomass harvest removals with fertilization (Nave et al. 2019). Forest land managers can also implement practices to increase SOC stocks by (i) reducing the extent of harvest area on a landscape level and lengthening the harvest cycle to allow long-term SOC accumulation, (ii) nitrogen (N)-fertilization, (iii) afforestation, (iv) reforestation of marginal croplands, and (v) introducing/favoring tree species that are faster-growing, more tolerant of heat or drought (Nave et al. 2019). As the climate gets drier in some regions, the SOC stocks especially in seasonally dry temperate coniferous forests may decrease in the long-term (Nave et al. 2019). This may also be the case for some temperate broadleaf and mixed forests. Otherwise, soil and land-use management practices to maintain and/or increase SOC stocks may be similar to those of temperate coniferous forests.

Globally, net GHG fluxes of tropical forests are comparable to those of boreal forests with both being smaller net C sinks than temperate forests (Harris et al. 2021). However, data for the tropical forest biome of the U.S., i.e., located in Hawaii and the U.S. Territories, are scanty (U.S. EPA 2021). Options to maintain SOC stocks in the U.S. tropical forest biome may include: (i) prevention of grass invasions that inhibit reforestation, (ii) agroforestry, (iii) fire management, (iv) maintaining landscape mosaic of forest fallows, (v) selection of high-SOC species for plantations, (vi) rotation lengths long enough to recover SOC stocks, (vii) mixed-species plantations, (viii) thinning to decrease fuels, (ix) reforestation of burned-over areas, and (x) grazer density control (Nave et al. 2019). Soil and land-use management practices to increase tropical forest biome SOC stocks in the U.S. may include: (i) reforestation, (ii) facilitation of N-fixer establishment, (iii) fertilization, (iv) soil erosion control, (v) selection of high-SOC species or genetic families for plantations, (v) retaining residues during salvage logging, (vi) reforestation of degraded soils and plantations with high-SOC species, and (vii) reforestation of eroded slopes (Nave et al. 2019).

Trees grow also outside of forest biomes in the U.S. (Crowther et al. 2015), and their SOC stocks may be affected by management. The net C sink in urban trees in U.S. settlements (0.13 Pg CO_2e y^{-1} [0.14 billion tn CO_2e y^{-1}]) was higher in 2019 than that in harvested wood products (0.11 Pg CO_2e y^{-1} [0.12 billion tn CO_2e y^{-1}]; U.S. EPA 2021). However, the U.S. does not estimate changes in SOC stocks for mineral soils in settlements remaining settlements including those under trees. There may be the potential of tree genus selection to maximize long-term SOC storage under urban trees but data are scanty (Edmondson et al. 2014). A collective management approach for urban SOC stocks including that affected by trees is needed (Lorenz and Lal 2015).

5.3 Terrestrial Biomes Under Agricultural Management

How management practices potentially affect SOC sequestration in crop- and grasslands in the U.S. will be summarized in the following sections. In 2019, temperate grassland soils in CONUS and Hawaii were an estimated net source of 15.1 Tg CO_2e (16.6 million tn CO_2e; U.S. EPA 2021), but there is a large discrepancy between modelling estimates and empirical data (Pendall et al. 2018). This uncertainty also affects the assessment of appropriate land management practices including whether moderate levels of grazing can maintain and/or increase SOC stocks (Pendall et al. 2018). Specifically, SOC sequestration rates in grazing lands are highly context-dependent (Godde et al. 2020). Thus, site-specific management of grazing intensity may contribute to higher SOC stocks in U.S. grasslands (Sollenberger et al. 2019). In theory, fertilization (N, P, K, micronutrients), irrigation, increasing species diversity, and sowing legumes and improved grass species may result in increases in temperate grassland SOC stocks (Conant et al. 2017). In the U.S., irrigation, seeding legumes and establishing high C input systems have been shown to increase grassland SOC stocks (Ogle et al. 2004). However, the SOC stock increase is temporal and will eventually cease when a new SOC stock equilibrium determined by climate and soil properties (i.e., composition of the mineral soil matrix, particle-size distribution, texture) is reached (Sollenberger et al. 2019). The grassland SOC stocks in the U.S. may also be altered by fire management but this is a debatable issue (Pellegrini et al. 2018).

In comparison to grasslands, croplands in the U.S. were established on higher quality soils supporting high plant biomass production but these soils have also historically lost larger amounts of SOC due to intensive soil disturbance (i.e., tillage), and removal of photosynthetically-fixed C with harvestable products together with low belowground C allocation (Sanderman et al. 2017). Thus, croplands can be specifically managed towards increases in SOC stocks until a new equilibrium determined by site factors is reached. Cropland soil C inputs can be increased by: (i) continuous plant cover all year (i.e., living roots) including cover crops in the off-season and perennial crops, (ii) growing crops with both high net primary production (NPP) and belowground C inputs, (iii) optimizing crop NPP by management of constraints such as competition from weeds, losses to herbivory, poor plant vigor, and suboptimal nutrient and water supply, and (iv) adding organic C with amendments (e.g., animal manure, compost, biochar). Another possibility to enhance cropland SOC sequestration is to control SOC losses. Management practices may include: (i) reducing disturbance by tillage (e.g., no-till [NT]) and perennialization including agroforestry, (ii) decreasing soil losses by water and wind erosion, and (iii) leaving non-harvestable biomass (e.g., crop residues) on site. Table 3.1 provides an overview on SOC sequestration rates in U.S. croplands. However, it is important to note that these are estimates based on the DayCent biogeochemical simulation model, and site-specific conditions are required for more detailed assessments of SOC dynamics on farms (Swan et al. 2015).

5.4 Enhancing the Adoption of Soil Organic Carbon Sequestration Practices

There is a wealth of knowledge available on how to reduce SOC losses from terrestrial biomes in the U.S., and on how to enhance SOC sequestration by implementing SOC-accruing soil and land-use management practices. Some examples were discussed in Chap. 3. Specifically, soils that have lost large amount of SOC in the past (carbon 'debt') such as cropland soils, and soils with unsaturated reactive mineral/metal-associated C pools can be managed for increases in SOC stocks. Soil conservation practices are relatively well understood, and are being currently deployed in production agriculture (e.g., cover cropping, intensified rotations, minimum tillage, advance nutrient management, integrated crop-livestock systems) but are not widely practiced and are in an early phase of adoption. Data on effects of soil and land-use management practices on SOC stocks are still limited, and, thus, the importance of SOC is often not considered in climate policy. Some argue that policy and economic incentives to increase SOC sequestration in the U.S. are weak while others report that extensive policies, programs, actions or targets related to SOC exist (Wiese et al. 2021; Amundson and Biardeau 2019). Incentives aimed at keeping C in terrestrial biomes and those adding C will ultimately also enhance SOC sequestration. The increased private and public sector interest in soil health which has accessory SOC sequestration benefits can also be leveraged to enhance the adoption of climate-smart and SOC-friendly practices in terrestrial biomes of the U.S. for climate, environmental and societal benefits. This increased recognition outside of the policy sphere is important as policies may change after an election cycle but climate change is here to stay into the foreseeable future as Earth's climate is committed to a warming trend by the amount of CO_2 already emitted into the atmosphere due to anthropogenic activities. However, a shift in presidential leadership alone cannot shift the politics, power dynamics, and paradigms that shape climate adaptation in the U.S. (Shi and Moser 2021). Nevertheless, national policies such as the Growing Climate Solutions Act (Braun et al. 2021), and the proposed carbon bank may support the adoption of protocols in the agricultural carbon market (Hammond et al. 2021). In contrast, the private sector's recognition of the reality of climate change prioritizes corporate investments, liability and profitability rather than human well-being or the inherent rights of nature (Bigger and Millington 2019). Thus, the long-term effects of the private sector's interest in climate adaptation on soil health and SOC sequestration in the U.S. is uncertain.

Scientists and policymakers in the U.S. are actively considering the political and technical feasibility of different climate solutions. Public policy can incentivize transition to negative carbon agriculture through payments for ecosystem services, agricultural product valuation methods that incorporate environmental footprint, insurance adjustments, lending/interest rates, grant support, and renewable energy generation credits (Northrup et al. 2021). Contributing to achieving negative emission agriculture is soil C storage which involves managing land, such as farm and grazing lands, forests, and wetlands, in ways that store increased amounts of C in the soil, thus keeping it out of the atmosphere (Sweet et al. 2021). However, the issue of climate change and the proposed policy solutions are politically polarized in the U.S. Important is public support, and a probability-based survey of the portion of the U.S. public that accepts the reality of climate change indicated that a majority expressed support for soil C storage as a climate change mitigation strategy, whether or not it involved biochar (Sweet et al. 2021). Specifically, soil C storage received more support than soil C storage with biochar (62% and 55%, respectively). Both soil C storage strategies trailed only afforestation and reforestation (AR) (73%) in terms of overall public support, and garnered more support than either bioenergy plus C capture and storage (BECCS) (32%) or direct air capture (DAC) (25%). Notably, this ordering of strategies in terms of policy support matched their ordering in terms of perceived naturalness. However, soil C storage with biochar was perceived by the public as being less natural than that without biochar. Thus, soil C storage that does not involve biochar may be especially politically feasible in the U.S. (Sweet et al. 2021).

The societal recognition of critical soil issues has very often involved the concept of 'soil quality' in the last three decades, and more recently that of 'soil health' (Baveye 2021). While the latter concept is discussed controversially (Janzen et al. 2021; Lehmann et al. 2020; Powlson 2020; Dick 2018), it is important to emphasize that soils of higher quality and more healthy soils are generally characterized by higher SOC stocks compared to those of lower quality and less health (Lal 2016). Adoption of emission negative farming practices (Lal 2021), such as conversion of plow till to conservation agriculture on erosion-prone or highly erodible lands, conserves soil and water and reduce erosion-induced lateral transport of SOC and emission of GHGs into the atmosphere (Lal 2020b).

5.4.1 *Federal Level*

The United States Department of Agriculture (USDA) may have a more prominent role in federal climate policy in the future (Bonnie et al. 2021). USDA's enormous and underappreciated discretionary financial resources and agency expertise may be particularly leveraged to reduce atmospheric GHGs through C sequestration and emissions reductions. Among key recommendations to achieve this is: (i) the establishment of a Carbon Bank using the Commodity Credit Corporation to finance large-scale investments in climate-smart land management practices; (ii) prioritization of climate-smart practices in implementation of Farm Bill conservation programs; and (iii) identifying opportunities to invest in natural infrastructure. Other recommendations include crop insurance, rural development grants and loans, and USDA procurement to incentivize climate-smart agriculture. Further, federal investment should be prioritized to address wildfire (Bonnie et al. 2021), which may have major effects on SOC sequestration and stocks in the future, especially in forest biomes of the Southwest and West.

On January 27, 2021, President Joe Biden signed Executive Order 14008 'Tackling the Climate Crisis at Home and Abroad' (The White House 2021). It acknowledges that America's farmers, ranchers, and forest landowners have an important role to play in combating the climate crisis and reducing GHG emissions, by sequestering C in soils, grasses, trees, and other vegetation and sourcing sustainable bioproducts and fuels. The U.S. Secretary of Agriculture was tasked to deliver a report with recommendations for a climate-smart agriculture and forestry (CSAF) strategy, and on how to encourage its voluntary adoption. The first CSAF report summarized initial conversations with Tribes and stakeholders across agriculture and forestry on how USDA should develop its CSAF strategy (USDA 2021). Partnerships with landowners, producers, state and local governments, Tribes, and other stakeholders across agriculture and forestry will be needed to advance CSAF. Several common themes emerged that need to be addressed to further SOC sequestration in terrestrial biomes. Among them is the identification of promising CSAF practices that deliver on climate outcomes, including GHG emission reductions, C sequestration, climate adaptation and resilience, and other co-benefits that are appropriate for large-scale adoption. The SOC measurement, monitoring and verification present challenges in the agricultural and forestry sectors due to variability among land types and practices, and the limited SOC data collection and testing, particularly regarding agricultural practices. Collection of SOC data is typically limited or rare, and the use of satellite and other newly available technologies inconsistent. Improved SOC data, however, is critical to help refine quantification tools and models for estimating and verifying SOC benefits of CSAF practices (USDA 2021).

Many USDA programs provide cost share and financial assistance for on-farm and forest conservation including funding and technical assistance for practices that have SOC benefits, including cover crops, precision agriculture, manure management, and forest restoration (USDA 2021). Examples of existing programs include Environmental Quality Incentives Program (EQIP), Conservation Stewardship Program (CSP), Agricultural Conservation Easement Program (ACEP), Conservation Reserve Program (CRP), Conservation Innovation Grants (CIG) and Regional Conservation Partnership Program (RCPP). SOC sequestration in forest biomes may be specifically benefit from Forest Legacy Program (FLP), Community Forest Program (CFP), Forest Stewardship Program (FSP), Sustainable Forestry African American Land Retention Program (SFLR), and Urban and Community Forestry (UCF) Program (USDA 2021). The National Sustainable Agriculture Coalition emphasizes that investments should be directed into these proven federal programs and not into carbon markets. Importantly, such program support would not be tied to a volatile carbon market, but be designed to encourage farmers to provide multiple environmental and natural resource benefits (NSAC 2021).

Regarding forest biomes, there is the potential to increase C sequestration capacity by 20% (0.19 Pg CO_2 [0.21 billion tn CO_2]) per year by facilitating re-plantings in understocked productive forestland (USDA 2021). This may also contribute to increases in forest SOC stocks. Major CSAF strategies targeting forests include: (i) increasing the rate of fuels reduction to decrease the risk of severe wildfire; (ii) increasing the rate of reforestation, especially after disturbances, and (iii) supporting applied forest research to inform climate mitigation and adaptation. Overall, major research efforts should be targeted towards the enhancement of soils for SOC storage in terrestrial biomes of the U.S. (USDA 2021).

To aid in moving forward initiatives to help farmers and ranchers improve the soil health resource base, the USDA Natural Resources Conservation Service (NRCS) has created a new Soil Health Division (SHD; Stott and Moebius-Clune 2017). Across NRCS regional boundaries, aims of SHD are to: (i) facilitate soil health technical training and education for stakeholders, (ii) work with partners to standardize soil health assessments, (iii) promote soil health management systems as part of the conservation planning process, and (iv) facilitate implementation and long-term adoption of soil health management systems on U.S. agricultural lands. It is anticipated that producers adopting soil health management systems will simultaneously address water quality and availability, habitat for biodiversity, and rural economic vitality, while sequestering C, adapting to and mitigating climate change, and feeding a growing population. Importantly, producers can achieve improvements in profits through better nutrient cycling, pest suppression, savings in energy and inputs, and enhanced water infiltration, storage, and drainage by adoption of soil health practices (Stott and Moebius-Clune 2017).

On June 24, 2021, the Growing Climate Solutions Act of 2021 passed the U.S. Senate (Braun et al. 2021). This bill authorizes USDA to establish a voluntary Greenhouse Gas Technical Assistance Provider and Third-Party Verifier Certification Program to help reduce entry barriers into voluntary environmental credit markets for farmers, ranchers, and private forest landowners. A voluntary environmental credit market is a market through which agriculture and forestry credits may be bought or sold. Entities eligible to participate in the program are: (i) providers of technical assistance to farmers, ranchers, or private forest landowners in carrying out sustainable land use management practices that prevent, reduce, or mitigate greenhouse gas emissions, or sequester C; or (ii) third-party verifiers that conduct the verification of the processes described in the protocols for voluntary environmental credit markets. USDA must publish: (i) a list of protocols and qualifications for eligible entities; (ii) information describing how entities may self-certify under the program; (iii) information describing how entities may obtain the expertise to meet the protocols and qualifications; and (iv) instructions and suggestions to assist farmers, ranchers, and private forest landowners in facilitating the development of agriculture or forestry credits and accessing voluntary environmental credit markets. This bill will enhance C sequestration and maintain and/or increase SOC stocks in terrestrial biomes of the U.S.

5.4.2 State Level

The importance of soil health and, thus, the SOC stock is increasingly recognized by decision makers at the state level. Soil health has recently gained significant attention from policymakers with much of the current policy discourse being driven by soil scientists and environmentalists, with little active engagement by economists (Stevens 2018). New research integrating soil science and economics is, thus, imperative for helping policymakers assess the benefits and costs of proposed soil health policies. As of August 2021, eighteen U.S. states have passed legislation on soil health (healthy-soils-legislation@googlegroups.com). While this looks promising towards enhancing SOC sequestration and stocks in U.S. biomes, soil health legislation may be rescinded after an election cycle, and it needs to be shown whether legislative support for soil health promoting activities will have a lasting impact on SOC sequestration and stocks. To date, adoption of soil health practices, in particular, inclusion of cover crops in rotations in the U.S. has been slow (LaRose and Myers 2019). In order to adopt soil-health practices, producers must (i) perceive negative outcomes from soil degradation, (ii) trust their sources of technical information, (iii) not face prohibitive costs or lost profits from implementing new practices, and (iv) believe that the new practices will produce some meaningful economic, environmental, or human health benefit (Stevens 2018).

5.4.3 Industry, Private and Public Sectors

There is a strong interest and market development for emission credits to meet the environmental targets of corporations and other large entities (Northrup et al. 2021). A few large companies in the U.S. have announced to purchase carbon credits in the near future. For example, Microsoft announced an agreement with Truterra, while IBM, JP Morgan Chase, Boston Consulting Group, Dogfish Head Craft Brewing, Shopify, Anheuser-Busch, and Barclays announced agreements with Indigo Ag (Plastina and Wongpiyabovorn 2021). Thus, the private-sector demand for C credits and commodities produced with CSAF practices may be an important lever for incentivizing CSAF practice adoption (USDA 2021). Consumers, processors and customers are increasingly demonstrating a preference for agricultural commodities produced using CSAF practices. Incentivizing CSAF practices may be based on: (i) voluntary markets for C where agriculture and forestry can provide C offsets or credits, (ii) sustainable supply chain initiatives, and (iii) "insetting" approaches where companies reduce GHG emissions within their own supply chains and production facilities. Carbon insets may be sold by farmers to downstream companies that use agricultural commodities in their supply chains (Plastina and Wongpiyabovorn 2021).

These markets can promote voluntary adoption of conservation technologies and practices, and leverage private-sector demand for GHG benefits associated with CSAF practices. However, there are many barriers including: (i) the relatively small scale of agricultural and small forestry offset projects; (ii) high transaction costs associated with project development, monitoring, reporting, and verification; and (iii) confusion in the C marketplace where there is a lack of consistency among approaches to protocols for generating GHG offsets from agriculture. As a result, only 2% of C offsets sold in the U.S. are generated from agricultural practices (USDA 2021). Except for livestock manure digestion projects, past efforts such as the CAR Nitrogen Management Protocol and California Compliance Offset Protocol for Rice Cultivation Projects have not resulted in large scale projects or significant credit volume (Hammond et al. 2021). Because of the many current uncertainties, companies with agricultural supply chains should only include GHG mitigation through SOC sequestration as part of their scope three reductions, i.e., a company's indirect emissions that occur in their value chain, including both upstream and downstream emissions (Oldfield et al. 2021). Companies can potentially make the greatest, most certain climate impact by prioritizing direct GHG emissions reductions (e.g., reduced nitrous oxide [N_2O] emissions via improved nutrient management and reduced CO_2 emissions via reduced tractor use), and avoid land conversions (Oldfield et al. 2021).

Carbon offset may describe the act of financing other climate change mitigation actions than those addressed by carbon credits to compensate or neutralize for the footprint of an organization (TSVCM 2021). In contrast, carbon credit may describe the verified GHG emissions reduction or removals generated, traded, retired. Carbon credits enable organizations to compensate or neutralize GHG emissions not yet eliminated by financing the avoidance/reduction of GHG emissions from other sources, or the removal of GHGs from the atmosphere. Projects generating carbon credits can be broadly grouped into GHG avoidance/reduction projects, such as renewable energy or avoided deforestation, and GHG removal/sequestration projects such as nature-based sequestration by reforestation and peatland restoration. The carbon credit sub-category avoided nature loss limits the loss of nature such as forests and peatlands that store and sequester C (TSVCM 2021).

Avoided nature loss is part of natural climate solutions or nature-based climate solutions (NCS) referring to the use of natural and working lands to mitigate human-caused climate change (Anderegg 2021). The carbon credit subcategory nature-based sequestration uses nature to sequester more C in the biosphere, including reforestation and restoring soil, mangroves, and peatlands. Nature-based sequestration is also part of NCS. In the short term, avoidance/reduction projects can and should be used, and existing nature loss projects should be financed and maintained for decades to come. However, a shift toward removal/sequestration is needed in the medium to long term for a century or longer. To function for climate mitigation, NCS efforts must meet criteria including: (i) providing additional C storage beyond what would have occurred otherwise, (ii) accounting for emissions that shift elsewhere due to NCS activities ("leakage"), (iii) having a net cooling effect on the climate by integrating biophysical and biogeochemical impacts of ecosystem changes, and (iv) achieving a level of C storage "permanence" over long periods of time (100 years or longer; Anderegg 2021). While practices that generate carbon offsets need to be maintained for long periods of time, practices that generate carbon insets might be only temporarily implemented (Plastina and Wongpiyabovorn 2021).

Buyers will have the opportunity to further delineate removal credits between geological C storage and biological C storage (TSVCM 2021). Importantly, high quality carbon credits, defined as being realistically baselined, additional, and permanent are required to increase confidence in carbon markets (Hammond et al. 2021). Carbon credits need to be rigorous, science-based, and transparent in their methods. Carbon credits are quantified as the difference in emissions that occurred in the project scenario relative to the baseline scenario (Hammond et al. 2021).

Corporations and entities may offset GHG emissions that are hard to abate (TSVCM 2021). This must be done through high-integrity carbon dioxide equivalent (CO_2eq) avoidance/reduction, and CO_2 removal/sequestration projects such that their compensation leads to genuine GHG emissions reductions and environmental benefits. For finance to flow to these projects, well-functioning voluntary carbon markets will be a critical enabler. A viable voluntary carbon market at scale could allow billions of dollars of capital to flow from those making commitments, such as carbon neutral or net-zero, into the hands of those with the ability to reduce and remove C. The market size in 2030 could be between $5 billion and up to over $50 billion assuming demand of 1–2 Gt CO_2 (1.1–2.2 billion tn CO_2). Voluntary carbon markets need to grow by more than 15-fold by 2030 in order to support the investment required to deliver the 1.5-degree pathway (TSVCM 2021). Voluntary carbon market incentives and rigorous protocols are already growing. There are financial mechanisms available to reward producers who implement land management practices that sequester additional carbon in soil and mitigate GHG emissions (Hammond et al. 2021).

Corporations, agricultural companies and other large entities in the U.S. voluntary committed to reducing their C footprint including making carbon-neutrality pledges which may also benefit SOC stocks within their resource supply chain (Lal 2020a). Many companies and organizations have committed to net zero carbon by 2040 (The Climate Pledge 2021), but this may enable business as usual in many cases. Also, the level of land use required to remove carbon may interfere with agriculture (Lal 2021). Instead, negative emission farming (NEF) should be supported by identifying technological options that increase agronomic productivity but also minimize GHG emissions (Lal 2021). In addition to SOC storage, a combination of innovations in digital agriculture, crop and microbial genetics, and electrification may contribute to achieve negative emissions from row-crop production (Northrup et al. 2021).

Companies in the U.S. such as Danone North America have also realized that investing to improve soil health is a business opportunity with the co-benefit of increasing SOC stores helping to deliver commitments to reduce emissions and limit global warming (WBCSD 2018). For example, the Midwest Row Crop Collaborative (MRCC) is a diverse coalition working to expand agricultural solutions that enhance soil health. Partners span the food supply chain in both the public and private sectors, including Bayer, Cargill, Environmental Defense Fund, General Mills, Kellogg Company, Land O'Lakes, McDonald's, PepsiCo, The Nature Conservancy, Unilever, Walmart and World Wildlife Fund. Until recently, MRCC supported a major initiative by the National Corn Growers Association, i.e., the Soil Health Partnership, a farmer-led initiative that promoted the adoption of soil health practices to ensure productive and sustainable agricultural systems (https://www.soilhealthpartnership.org/). Over 200 farms across 16 midwestern U.S. states conducted research on cover cropping, nutrient management, and conservation tillage. SHP supported farmers in better understanding on how to potentially mitigate some of the effects of climate change by promoting practices that sequester SOC. Initial results indicated that cover cropping had positive impact on some soil health indicators but did not change soil organic matter (g kg^{-1}) in 0–15 cm (0–6 in) depth with duration of cover crop adoption ranging on average between 1.3 and 2.2 years (Wood and Bowman 2021). Also, soil health indicators across control and soil health treatments often did not directly relate to corn (*Zea mays* L.) and soybean (*Glycine max* L.) crop yields (Crookston et al. 2021). Thus, Wood and Bowman (2021) suggested that there is opportunity and need for further assessment of how real-world use of soil health practices translate into tangible agronomic and environmental targets. However, after seven years of activities, SHP closed its doors in May 2021.

Some U.S. companies such as Danone North America, Indigo Agriculture and General Mills are incentivizing the adoption of regenerative agriculture practices with potential co-benefits for SOC sequestration (Renton et al. 2020). There is no common definition for regenerative agriculture but descriptions are based on processes (e.g., use of cover crops, the integration of livestock, and reducing or eliminating tillage), outcomes (e.g., to improve soil health, to sequester C, and to increase biodiversity), or combinations of the two (Newton et al. 2020).

5.4.4 Carbon Farming and Carbon Markets

Using land for carbon removal or carbon farming can be described as the management of C pools, flows and GHG fluxes at farm level, with the purpose of mitigating climate change (COWI, Ecologic Institute and IEEP 2021). Carbon farming involves the management of both land and livestock, all pools of C in soils, material and vegetation plus fluxes of CO_2, methane (CH_4) and N_2O. Carbon farming includes C removal from the atmosphere, avoided GHG emissions and emission reductions from ongoing agricultural practices (COWI, Ecologic Institute and IEEP 2021). Carbon farming can provide additional financial, social, and environmental co-benefits that provide different levels of private and public net benefit (Lin et al. 2013). It may include designing crops with the attributes: (i) increased belowground C allocation for larger and deeper root biomass; (ii) interactions with a tailored, synthetic soil microbiome for increased rhizosphere sink strength and enhanced plant growth-promoting properties that facilitate nutrient acquisition and water-use efficiency; and (iii) increased source strength for enhanced photosynthesis and biomass accumulation (Jansson et al. 2021).

5.4.4.1 Voluntary Initiatives

The adoption of carbon farming practices can be supported by incentivizing farmers, growers and ranchers towards participation in carbon markets. Appropriate standards and adequate supply and demand are essential for successful carbon markets. Market payments can be generated via outcomes verified at the field level (Bruner and Brokish 2021). A few active and pilot primarily voluntary markets exist in the U.S. for carbon sequestration and GHG emission reductions. Carbon offsets are generated as a measurable reduction of GHG emissions from an activity or project in one location that is used to compensate for carbon emissions occurring elsewhere, e.g., by selling them as carbon credits. Among carbon market entities active in the U.S. are Nori, Indigo Ag, Soil and Water Outcomes Fund, and Ecosystem Services Market Consortium. The requirements for implementation of practices and estimations of outcomes verified at the field level for generating market payments differ among them. Oldfield et al. (2021) compared eight U.S. protocols with a range of quantification activities, structural considerations and requirements intended to ensure the integrity of quantified credits. For example, Nori requires implementation of a new practice, with a look-back of up to 5 years during the pilot phase. This, however, calls into question the premise of additionality for these pilot phase credits, and most other carbon programs do not allow for these "look-back" periods (Oldfield et al. 2021). Estimation of outcomes by Nori is based on a soil sample reference network-based modeling with cost incurred by Nori. In addition, farmers have the option to validate data via soil sampling with farmers incurring sampling cost (Bruner and Brokish 2021).

Indigo Ag does require implementation of a new practice with a look-back of 2 growing seasons (Bruner and Brokish 2021). Modeling (biogeochemical and statistical) plus soil sampling is the basis for estimation of outcomes. Indigo Ag assumes cost, and does not charge growers for anything. Recently, Indigo Ag proposed a protocol that may enable scalable, high-quality credits through four main advances: (i) allowing flexibility in the use of biogeochemical models that meet explicit performance requirements, (ii) enabling a new approach to field-level, modeled baselines, (iii) supporting a hybrid approach of credit generation using both soil measurement and modeling, and (iv) requiring a new type of credit uncertainty quantification that accounts for multiple sources of uncertainty (Hammond et al. 2021). These recent advances may support agricultural credit quantification that enables payments to offset transitional costs for growers. This may be achieved at large enough scales to create a robust market, with a level of rigor that ensures any credited emission reductions have real climate impact. Further, innovations in soil analyses, advances in research, and improvements in data collection may improve the potential for agricultural carbon credits to scale (Hammond et al. 2021).

The Soil and Water Outcomes Fund requires a new practice for project enrollment (Bruner and Brokish 2021). The fund bases its estimates on modeling, with 10% of fields subject to in-field soil and water sampling at no cost to the farmer. The Ecosystem Services Market Consortium requires also a new practice, but investigates potential of payments to producers already implementing conservation practices. A peer reviewed biogeochemical model plus soil sampling are used for estimation of outcomes. The consortium assumes costs and includes in asset price to buyers (Bruner and Brokish 2021).

The Ag Web Farm Journal compared additional carbon market programs (Farm Journal Editors 2021). For example, Bayer's carbon initiative pays producers for adopting climate-smart practices such as no-till, strip-till and the planting of cover crops. Producers are required to plant corn or soybeans, have an active FieldView Plus account, and agree to share the data needed for the program. The CIBO Impact program is called the REAP program: Rapid Enrollment, Annual Payment. CIBO verifies regenerative practices through remote sensing and interviews. Subsequently, a farmer's regenerative potential becomes a CIBO carbon credit that is available for sale on the CIBO marketplace. The grower retains ownership of the credits. For the Farmer Business Network program, producers share information with Gradable on their crop production practices, including planting, fertilizer applications, tillage and harvest. The information is processed with artificial intelligence that leverages 240 million acre (97 Mha)-events of farm data from the Farm Business Network. Gradable validates and distills the practices into a single farm-level score, which allows farms to be rewarded for practices without having to share detailed practice information with buyers. Under development is the voluntary Nutrien program activating through Nutrien Ag Solutions. It will involve a producer program participation agreement, which requires producers to fulfill sustainable farming practice obligations to receive grower payments. Producers participating in the program will agree to start providing data and implementing practice changes shortly after signing up. TruCarbon is activating through Truterra and its 24 retail partners within the larger WinField United network. The program helps farmers generate and sell carbon credits to private-sector buyers. This is a look-back program for carbon farmers who have already sequestered carbon – up to 5 years ago (Farm Journal Editors 2021).

Carbon credits estimated based on models are associated with uncertainties as some models are in the early phases and lack many of the large-scale training datasets (Abram 2021). Modeling to assess changes in soil carbon as a result of changing management refers to the use of a process-based model or a biogeochemical model, which requires inputs on weather, soil, and management history to simulate soil processes that alter soil C cycling and storage (Baltensperger et al. 2021). There are no clear protocols for quantifying uncertainty. Also, some models lack flexibility if a new technology comes along. However, with the interest in soil C there will likely be a lot of innovation in technologies to increase soil C in the next 5–10 years. That's where physical measurement has advantages – creating a baseline, allowing flexibility for changes in practices and generating a high degree of trust in the carbon credits (Abram 2021). Measurement-based assessment includes traditional soil coring, non-destructive proximal soil sampling, and remote sensing (Baltensperger et al. 2021). Similar to modeling, physical measurements are associated with uncertainty limitations. This includes cost of sampling for soil coring, measurement precision for proximal sensing, and extrapolation across management histories for remote sensing. However, accurate quantification of soil C is critical to the integrity and credibility of soil carbon credits in carbon markets. The ability to quantify uncertainty in estimating or measuring agricultural GHG emissions and changes is important for market-based accounting. Because soil C stocks are at risk of losses or reversals from storage, market standards require that they be monitored and replaced if transacted credits representing increased soil C are lost (Baltensperger et al. 2021).

To sum up, voluntary agricultural carbon markets are an emerging field with most of them not fully launched (Plastina and Wongpiyabovorn 2021). The agriculture carbon credits market can be currently characterized as an unarticulated patch of coexisting programs with different rules, incentives, and penalties (Plastina and Wongpiyabovorn 2021). Whether SOC sequestration and stocks of croplands and grasslands will benefit from them is not known. Only few carbon offsets sold in the U.S. are currently generated from agricultural practices (USDA 2021). However, carbon markets have enormous potential to incentivize and reward climate progress but those must be paired with a strong regulatory backing (Oldfield et al. 2021).

5.4.4.2 Compliance Programs

Compared to the voluntary primarily agricultural carbon market, better developed is the forest carbon market. California is the only U.S. state with a compliance market, i.e., an economy-wide cap-and-trade program (Hamrick and Gallant 2017). Until 2021, entities covered under the cap-and-trade were allowed to use offsets for up to

8% of their compliance obligation, although that will fall to 4% in 2021–2025, then increase to 6% in 2026–2030. California has three operational forestry and land-use carbon methodologies: U.S. forestry, urban forestry, and rice (*Oryza sativa* L.) cultivation, although no credits have yet been issued under rice cultivation or urban forestry protocols. In fact, forest carbon offsets are currently widely used as part of the cap-and-trade carbon market and the State's Natural and Working Lands Climate Change Implementation Plan (Anderegg 2021). Carbon offsets are also issued on the voluntary carbon market. All offset projects of the California market must be based in CONUS (Hamrick and Gallant 2017). In 2014, California's program linked with the Canadian province of Québec. Coffield et al. (2021) indicated that the climate risks to forest offset projects located in California are remarkably high, and substantially underestimated in current versions of voluntary and compliance markets' forest offset protocols. Thus, NCS efforts are most likely to be potentially useful for climate mitigation if they are paired with aggressive reductions of GHG emissions from human sources (Anderegg 2021).

The Regional Greenhouse Gas Initiative (RGGI) was the first compliance-based market for reducing GHG emissions in the U.S. (Hamrick and Gallant 2017). The program limits emissions from fossil fuel power plants of 25 megawatts and larger within nine northeastern states: Connecticut, Delaware, Maine, Maryland, Massachusetts, New Hampshire, New York, Rhode Island, and Vermont. RGGI only allows the use of offsets if allowance prices exceed a certain level. RGGI permits several types of offsets, including offsets from U.S. forest projects (reforestation, improved forest management, avoided conversion). However, afforestation would only be allowed in Connecticut and New York (Hamrick and Gallant 2017).

Overall, the engagement of farmers, growers and ranchers in the agricultural carbon market has been slow as the community is suspicious that cap and trade programs such as those in California and RGGI are simply a pathway to additional regulation (Toor et al. 2021). None of these programs allow row crop agriculture as a source of carbon offsets. Sometimes complex transaction costs may also limit interest by farmers, growers and ranchers to participate in the markets, especially where they require disclosure of practice history, inputs, and soil types. Suspicion of the intent of these government-controlled markets in an environment of increasing agricultural regulations often prevents engagement (Toor et al. 2021). Nevertheless, interest in agriculture's role in carbon markets is increasing but also scrutiny on how to accurately measure, report, and verify (MRV) changes in GHG emissions from agriculture, including for soil carbon sequestration.

The integrity of soil carbon credits depends on MRV. Oldfield et al. (2021) compared twelve publicly available SOC MRV protocols published by nonprofit carbon registries and by private carbon crediting marketplaces with eight of the protocols from the U.S. Some of the protocols use soil sampling only, some combine sampling with process-based modeling, and others use only modeling and remote sensing. There is little evidence that existing models can accurately capture SOC change at the field level under all management interventions for all combinations of soils and climate. However, in the short-term, carbon credits will primarily be issued based on modeled results. Designing an effective soil sampling strategy that adequately captures spatial heterogeneity and reduces uncertainty in SOC stock estimates is another critical issue. To address the many challenges and uncertainties, the different MRV protocols should be compared to help determine the degree to which those equivalently account for net GHG reductions. Grouping together multiple farm-scale projects is recommended to account for additionality and leakage, reduce risks of reversal, help provide MRV cost savings, and support participation of diverse farm operations within any crediting program. High-quality, open-access datasets for model calibration, benchmarking, and baseline and additionality determination should be developed. Finally, the continued development of cost-effective approaches to MRV should be supported using emerging technology to help produce accurate and scalable solutions for quantifying net GHG reductions. Importantly, most of the protocols have not been adopted while voluntary carbon markets are developing at a rapid pace (Oldfield et al. 2021).

5.4.5 *Carbon Pricing*

The SOC stock has monetary value for U.S. society, and, thus, land managers and owners should be rewarded financially for maintaining and/or increasing SOC stocks. As discussed in Chap. 2, to limit global average warming to 1.5 °C (2.7 °F), a social cost of carbon (SCC) price of USD 100 per Mg (USD 91 per ton) of CO_2, and possibly as high as USD 200 per Mg (USD 182 per ton) of CO_2 in 2020 U.S. dollars should be established by 2030 (Wagner 2021). To date, carbon pricing has had a limited impact on emissions, jeopardizing the aim to limit warming (Green 2021). However, even carbon pricing policies with broad coverage such as those in California lack extensive independent evaluations. There is also very little price transparency in these carbon markets (Thompson et al. 2021).

The value of SOC can be assessed based on the avoided SCC internalizing the long-term environmental and health damage resulting from GHG emissions (Mikhailova et al. 2019). The avoided SCC has been estimated at USD 42 per Mg (USD 46 per ton) of CO_2 in 2007 U.S. dollars. Thus, producers should be rewarded by fair pricing to facilitate the implementation of SOC-accruing practices in agriculture and forestry (USDA 2021). However, costs of carbon credits are low and represents the primary reason for the low engagement of farmers and ranchers in carbon markets. Further, carbon prices simply are also not high enough to generate substantial emissions reductions (Green 2021). Low prices are pervasive, i.e., the vast majority of carbon prices are well below even the most conservative estimates of the SCC. Recent global estimates for the SCC ranged between USD 80 and USD 300 per Mg (USD 73 and USD 272 per ton) CO_2 (Pindyck 2019). As of 2019, existing carbon pricing schemes only cover about 20% of global emissions, and more than two-thirds of these have prices below USD 20 per Mg (USD 18 per ton) of CO_2eq (Rosenbloom et al. 2020). This pricing is far too low to be effective, and increasing coverage and prices presents serious challenges.

By a fair price on C or CO_2, implementation of SOC sequestering practices by forest landowners, farmers, growers and ranchers may be incentivized. Economy-wide pricing of C at around 100 USD per Mg CO_2 (91 USD per tn CO_2) appears to make a number NCS cost-effective with some of them also including soil C sequestration (Griscom et al. 2017). However, economy-wide carbon pricing alone might be insufficient to result in implementation of NCS (Anderegg 2021; Nolan et al. 2021). Specifically, agriculture and soil-carbon-based NCS interventions have unique challenges relating to the small scales of deployment in contrast to forest-based solutions. Otherwise, implementation in agriculture may be motivated by increased agronomic yield or other valued co-benefits on short timescales, while the increased C storage accumulates over longer timescales. However, soil-carbon-based projects pose significant measurement and verification challenges that feed back into cost-efficiency, i.e., the costs for quantifying changes in soil C stocks are very large relative to the value of the stock changes at reasonable carbon prices. Further, mechanisms to manage the risks of leakage and reversal cut into economic and financial viability, potentially enough to make many projects non-viable. Among key near-term barriers for robust implementation of NCS is creating rigorous and workable governance mechanisms. These mechanisms are central to investments that can increase C storage in the terrestrial biosphere (Nolan et al. 2021).

Conflicts over carbon pricing in the U.S. are intense (Green 2021). There is a long history, spanning from repeated failures at the federal level, to a mix of success and failure at the state level. In 1993, President Clinton proposed an energy tax (i.e., the BTU tax), which died in the Senate after considerable opposition from both Republicans and Democrats. Subsequent efforts to create a national cap-and-trade scheme have also failed. As discussed previously, there has been more success in creating emissions trading schemes, i.e., California and the Northeastern states in RGGI have had emissions trading in effect since 2012 and 2009, respectively. However, carbon taxes remain absent from U.S. state policy (Green 2021).

Importantly, some have suggested that carbon pricing may not be sufficient to mitigate climate change (Rosenbloom et al. 2020). Specifically, carbon pricing may have weaknesses with regard to five central dimensions: (i) problem framing and solution orientation, (ii) policy priorities, (iii) innovation approach, (iv) contextual considerations, and (v) politics. Carbon pricing functions poor in sectors with small, diffuse-point sources. For example, agrofood systems are characterized by manifold commodities, dispersed production (millions of farmers) in highly variable contexts (soil conditions, climate, local communities), and deeply entrenched cultural conventions, such as tastes and dietary habits. This all makes it extremely difficult to assess the level of an effective carbon price and implement this throughout the agrofood system. Instead of focusing on carbon pricing, Rosenbloom et al. (2020) proposed a sustainability transition policy to move beyond market failure reasoning and focus on fundamental changes in existing sociotechnical systems such as energy, mobility, food, and industrial production.

An alternative approach to incentivize the adoption of SOC-sequestering practices may be higher price premiums for products originating from these practices. This support scheme may be designed similar to that for organic agriculture (OA).

The adoption of OA practices in the U.S. is facilitated by price premiums for OA products. Driven by increasing consumer demand, the United States Department of Agriculture (USDA) has launched an organic certification scheme (Wiggins and Nandwani 2020). The certified OA production area in the U.S. has increased steadily since the inception of the 1990 Organic Foods Production Act. The National Organic Program (NOP) established in 2000 restricts the use of the term "organic" to certified organic producers. Consumers can choose between conventional and more pricy organic products that are clearly identified by an official USDA seal. This higher price can be understood as a kind of tax but with the advantage of directly rewarding farmers, growers and ranchers for using certified OA practices. Another advantage of this USDA certification scheme is that it is broadly supported by society, and independent of policies and election outcomes. U.S. consumers and the public and private sector are increasingly concerned with the environmental footprint of food and other agricultural and forest products, and a 'climate-smart' or 'soil organic carbon-friendly' seal may support consumer choices towards products contributing to climate change adaptation and mitigation (Dietz 2014). There is some evidence from other countries that citizens are willing to pay to support soil carbon sequestration programs (Kragt et al. 2016; Glenk and Colombo 2011). The success of the USDA organic certification over decades is promising as a similar carbon certification scheme needs to be established for decades and longer to address the climate crisis. However, whether such a certification scheme results in increased adoption of carbon farming practices needs to be studied.

5.5 Conclusions

Terrestrial biomes in the U.S. can be managed for C sequestration, and maintaining and enhancing SOC stocks. This has numerous environmental and societal benefits and land managers must be rewarded for this service to society. However, many barriers to the adoption of climate-smart and SOC-friendly practices exist including site-specific knowledge about the most effective and economically viable soil and land-use management practices. Federal and state policies can incentivize the adoption of these innovative practices but legislation may be rescinded or revised after a policy change. The nascent voluntary carbon market may be another option to enhance SOC stocks as the private sector is increasingly looking for options to offset supply chain GHG emissions by buying carbon credits with many corporations committing to net zero emissions. Many barriers towards including SOC in carbon markets remain such as monitoring, reporting and verification of changes in SOC stocks, leakage and permanence, and the low CO_2 prices relative to the investments needed to implement and maintain SOC-friendly practices. To build trust and confidence, standardization of soil carbon credits is needed. Price premiums on climate-smart and SOC-friendly products signified by an official seal on products may be another possibility similar to the successful USDA seal for products from organic agriculture. Such a payment scheme for SOC stewardship would directly reward farmers, growers, ranchers and forest landowners for maintaining and increasing SOC stocks.

References

Abram M (2021) Why soil measurement is a key challenge for carbon farming. Farmers Weekly, 10 June 2021. https://www.fwi.co.uk/arable/land-preparation/soils/why-soil-measurement-is-a-key-challenge-for-carbon-farming

Amundson R, Biardeau L (2019) Opinion: soil carbon sequestration is an elusive climate mitigation tool. Proc Natl Acad Sci USA 116:11652–11656. https://doi.org/10.1073/pnas.1908917116

Anderegg WRL (2021) Gambling with the climate: how risky of a bet are natural climate solutions? AGU Adv 2:e2021AV000490. https://doi.org/10.1029/2021AV000490

Baltensperger D, Basu M, Dhou Z et al (2021) The role of agricultural science and technology in Climate 21 Project implementation. CAST Commentary QTA2021-1 June 2021. https://www.cast-science.org/publication/the-role-of-agricultural-science-and-technology-in-climate-21-project-implementation/

Baveye PC (2021) Soil health at a crossroad. Soil Use Manag 37:215–219. https://doi.org/10.1111/sum.12703

Bigelow DP, Borchers A (2017) Major uses of land in the United States, 2012. EIB-178, U.S. Department of Agriculture, Economic Research Service

Bigger P, Millington N (2019) Getting soaked? Climate crisis, adaptation finance, and racialized austerity. Environ Plan E Nat Space 3:601–623. https://doi.org/10.1177/2514848619876539

Bonnie R, Jones L, Harrell M (2021) Climate 21 project United States Department of Agriculture (USDA) transition memo. https://climate21.org/documents/C21_USDA.pdf

Braun M, Stabenow D, Graham L, Whitehouse S (2021) "Growing Climate Solutions Act of 2021." Senate Bill S.1251. https://www.congress.gov/bill/117th-congress/senate-bill/1251

Bruner E, Brokish J (2021) Ecosystem market information: background and comparison table [Fact sheet]. Illinois Sustainable Ag Partnership. https://ilsustainableag.org/wp-content/uploads/2021/02/EcosystemMarketInformation.pdf

Chambers A, Lal R, Paustian K (2016) Soil carbon sequestration potential of US croplands and grasslands: implementing the 4 per thousand initiative. J Soil Water Conserv 71:68A–74A. https://doi.org/10.2489/jswc.71.3.68A

Chen J, Elsgaard L, van Groenigen KJ et al (2020) Soil carbon loss with warming: new evidence from carbon-degrading enzymes. Glob Chang Biol 26:1944–1952. https://doi.org/10.1111/gcb.14986

Coffield SR, Hemes KS, Koven CD, Goulden ML, Randerson JT (2021) Climate-driven limits to future carbon storage in California's wildland ecosystems. AGU Adv 2:e2021AV000384. https://doi.org/10.1029/2021AV000384

Conant RT, Cerri CEP, Osborne BB, Paustian K (2017) Grassland management impacts on soil carbon stocks: a new synthesis. Ecol Appl 27:662–668. https://doi.org/10.1002/eap.1473

COWI, Ecologic Institute and IEEP (2021) Technical Guidance Handbook – setting up and implementing result-based carbon farming mechanisms in the EU Report to the European Commission, DG Climate Action, under Contract No. CLIMA/C.3/ETU/2018/007. COWI, Kongens Lyngby

Crookston BS, Yost MA, Bowman M et al (2021) Soil health spatial-temporal variation influence soil security on Midwestern, U.S. farms. Soil Secur 3:100005. https://doi.org/10.1016/j.soisec.2021.100005

Crowther TW, Glick HB, Covey KR et al (2015) Mapping tree density at a global scale. Nature 525:201–205. https://doi.org/10.1038/nature14967

Dick R (2018) Soil health: the theory of everything (terrestrial) or just another buzzword? Crop, Soils Agron News 63:12–17. https://doi.org/10.2134/csa2018.63.1114
Dietz W (2014) Understanding environmentally significant consumption. Proc Natl Acad Sci U S A 111:5067–5068. www.pnas.org/cgi/doi/10.1073/pnas.1403169111
Domke G, Williams CA, Birdsey R et al (2018) Chapter 9: forests. In: Cavallaro N, Shrestha G, Birdsey R et al (eds) Second state of the carbon cycle report (SOCCR2): a sustained assessment report. U.S. Global Change Research Program, Washington, DC, pp 365–398. https://doi.org/10.7930/SOCCR2.2018.Ch9
Edmondson JL, O'Sullivan OS, Inger R et al (2014) Urban tree effects on soil organic carbon. PLoS One 9:e101872. https://doi.org/10.1371/journal.pone.0101872
Farm Journal Editors (2021) Clarity on carbon's potential: compare nine of the leading markets. Ag Web Farm Journal. https://www.agweb.com/news/business/technology/clarity-carbons-potential-compare-nine-leading-markets
Glenk K, Colombo S (2011) Designing policies to mitigate the agricultural contribution to climate change: an assessment of soil based carbon sequestration and its ancillary effects. Clim Chang 105:43–66. https://doi.org/10.1007/s10584-010-9885-7
Godde CM, de Boer IJM, zu Ermgassen E et al (2020) Soil carbon sequestration in grazing systems: managing expectations. Clim Chang 161:385–391. https://doi.org/10.1007/s10584-020-02673-x
Green JF (2021) Does carbon pricing reduce emissions? A review of expost analyses. Environ Res Lett 16:043004. https://doi.org/10.1088/1748-9326/abdae9
Green JK, Seneviratne SI, Berg AM et al (2019) Large influence of soil moisture on long-term terrestrial carbon uptake. Nature 565:476–479. https://doi.org/10.1038/s41586-018-0848-x
Griscom BW, Adams J, Ellis PW et al (2017) Natural climate solutions. Proc Natl Acad Sci U S A 114:11645–11650. www.pnas.org/cgi/doi/10.1073/pnas.1710465114
Hammond AAJ, Motew M, Brummitt CD et al (2021) Implementing the soil enrichment protocol at scale: opportunities for an agricultural carbon market. Front Clim 3:686440. https://doi.org/10.3389/fclim.2021.686440
Hamrick K, Gallant M (2017) Fertile ground: state of forest carbon finance 2017. Forest Trends' Ecosystem Marketplace. https://www.forest-trends.org/publications/fertile-ground/
Harris NL, Gibbs DA, Baccini A et al (2021) Global maps of twenty-first century forest carbon fluxes. Nat Clim Chang 11:234–240. https://doi.org/10.1038/s41558-020-00976-6
Hayes DJ, Vargas R, Alin SR et al (2018) Chapter 2: the North American carbon budget. In: Cavallaro N, Shrestha G, Birdsey R et al (eds) Second state of the carbon cycle report (SOCCR2): a sustained assessment report. U.S. Global Change Research Program, Washington, DC, pp 71–108. https://doi.org/10.7930/SOCCR2.2018.Ch2
Hibbard KA, Hoffman FM, Huntzinger D, West TO (2017) Changes in land cover and terrestrial biogeochemistry. In: Wuebbles DJ, Fahey DW, Hibbard KA et al (eds) Climate science special report: fourth national climate assessment, vol I. U.S. Global Change Research Program, Washington, DC, pp 277–302. https://doi.org/10.7930/J0416V6X
Homer C, Dewitz J, Jin S et al (2020) Conterminous United States land cover change patterns 2001–2016 from the 2016 National Land Cover Database. ISPRS J Photogramm 162:184–199. https://doi.org/10.1016/j.isprsjprs.2020.02.019
Hristov AN, Johnson JMF, Rice CW et al (2018) Chapter 5: agriculture. In: Cavallaro N, Shrestha G, Birdsey R et al (eds) Second state of the carbon cycle report (SOCCR2): a sustained assessment report. U.S. Global Change Research Program, Washington, DC, pp 229–263. https://doi.org/10.7930/SOCCR2.2018.Ch5
Jansson C, Faiola C, Wingler A et al (2021) Crops for carbon farming. Front Plant Sci 12:636709. https://doi.org/10.3389/fpls.2021.636709
Janzen HH, Janzen DW, Gregorich EG (2021) The 'soil health' metaphor: illuminating or illusory? Soil Biol Biochem 159:108167. https://doi.org/10.1016/j.soilbio.2021.108167
Kolka R, Trettin C, Tang W et al (2018) Chapter 13: terrestrial wetlands. In: Cavallaro N, Shrestha G, Birdsey R et al (eds) Second state of the carbon cycle report (SOCCR2): a sustained assess-

ment report. U.S. Global Change Research Program, Washington, DC, pp 507–567. https://doi.org/10.7930/SOCCR2.2018.Ch13

Kragt ME, Gibson FL, Maseyk F, Wilson KA (2016) Public willingness to pay for carbon farming and its co-benefits. Ecol Econ 126:125–131. https://doi.org/10.1016/j.ecolecon.2016.02.018

Lal R (2016) Soil health and carbon management. Food Energy Secur 5:212–222. https://doi.org/10.1002/fes3.96

Lal R (2020a) The role of industry and the private sector in promoting the "4 per 1000" initiative and other negative emission technologies. Geoderma 378:114613. https://doi.org/10.1016/j.geoderma.2020.114613

Lal R (2020b) Regenerative agriculture for food and climate. J Soil Water Conserv 75:123A–124A. https://doi.org/10.2489/jswc.2020.0620A

Lal R (2021) Negative emission farming. J Soil Water Conserv 76:61A–64A. https://doi.org/10.2489/jswc.2021.0419A

LaRose J, Myers R (2019) Adoption of soil health systems based on data from the 2017 U.S. Census of Agriculture. Soil Health Institute. https://soilhealthinstitute.org/wp-content/uploads/2019/07/Soil-Health-Census-Report.pdf

Lehmann J, Bossio DA, Kögel-Knabner I, Rillig MC (2020) The concept and future prospects of soil health. Nat Rev Earth Environ 1:544–553. https://doi.org/10.1038/s43017-020-0080

Lin BB, Macfadyen S, Renwick AR, Cunningham SA, Schellhorn NA (2013) Maximizing the environmental benefits of carbon farming through ecosystem service delivery. BioSci 63:793–803. https://doi.org/10.1525/bio.2013.63.10.6

Lorenz K, Lal R (2015) Managing soil carbon stocks to enhance the resilience of urban ecosystems. Carbon Manage 6:35–50. https://doi.org/10.1080/17583004.2015.1071182

Luo Z, Rossel RAV, Shi Z (2020) Distinct controls over the temporal dynamics of soil carbon fractions after land use change. Glob Chang Biol 26:4614–4625. https://doi.org/10.1111/gcb.15157

Mikhailova EA, Groshans GR, Post CJ, Schlautman MA, Post GC (2019) Valuation of soil organic carbon stocks in the contiguous United States based on the avoided social cost of carbon emissions. Resources 8:153. https://doi.org/10.3390/resources8030153

National Academies of Sciences, Engineering, and Medicine (2019) Negative emissions technologies and reliable sequestration: a research agenda. The National Academies Press, Washington, DC. https://doi.org/10.17226/25259NAS

National Sustainable Agriculture Coalition (2021) Invest in climate solutions for farmers. https://sustainableagriculture.net/blog/nsac-report-on-climate-solutions-for-farmers-invest-in-proven-federal-programs/

Nave L, Marín-Spiotta E, Ontl T, Peters M, Swanston C (2019) Soil carbon management. In: Busse M, Giardina CP, Morris DM, Page-Dumroese DS (eds) Global change and forest soils, Developments in soil science 36. Elsevier, Amsterdam, pp 215–257. https://doi.org/10.1016/B978-0-444-63998-1.00011-2

Newton P, Civita N, Frankel-Goldwater L, Bartel K, Johns C (2020) What is regenerative agriculture? A review of scholar and practitioner definitions based on processes and outcomes. Front Sustain Food Syst 4:577723. https://doi.org/10.3389/fsufs.2020.577723

Nolan CJ, Field CB, Mach KJ (2021) Constraints and enablers for increasing carbon storage in the terrestrial biosphere. Nat Earth Environ Rev 2:436–446. https://doi.org/10.1038/s43017-021-00166-8

Northrup DL, Basso B, Wang MQ, Morgan CLS, Benfey PN (2021) Novel technologies for emission reduction complement conservation agriculture to achieve negative emissions from row-crop production. Proc Natl Acad Sci USA 118:e2022666118. https://doi.org/10.1073/pnas.2022666118

Ogle SM, Conant RT, Paustian K (2004) Deriving grassland management factors for a carbon accounting method developed by the intergovernmental panel on climate change. Environ Manag 33:474–484. https://doi.org/10.1007/s00267-003-9105-6

Oldfield EE, Eagle AJ, Rubin RL et al (2021) Agricultural soil carbon credits: making sense of protocols for carbon sequestration and net greenhouse gas removals. Environmental

Defense Fund, New York. edf.org/sites/default/files/content/agricultural-soil-carbon-credits-protocolsynthesis.pdf

Park A, Puettmann K, Wilson E et al (2014) Can boreal and temperate forest management be adapted to the uncertainties of 21st century climate change? Crit Rev Plant Sci 33:251–285. https://doi.org/10.1080/07352689.2014.858956

Pellegrini AFA, Ahlström A, Hobbie SE et al (2018) Fire frequency drives decadal changes in soil carbon and nitrogen and ecosystem productivity. Nature 553:194–198. https://doi.org/10.1038/nature24668

Pendall E, Bachelet D, Conant RT et al (2018) Chapter 10: grasslands. In: Cavallaro N, Shrestha G, Birdsey R et al (eds) Second state of the carbon cycle report (SOCCR2): a sustained assessment report. U.S. Global Change Research Program, Washington, DC, pp 399–427. https://doi.org/10.7930/SOCCR2.2018.Ch10

Pindyck RS (2019) The social cost of carbon revisited. J Environ Econ Manag 94:140–160. https://doi.org/10.1016/j.jeem.2019.02.003

Plastina A, Wongpiyabovorn O (2021) How to grow and sell carbon credits in US agriculture. Ag Decision Maker File A1–76. extension.iastate.edu/agdm

Powlson DS (2020) Soil health – useful terminology for communication or meaningless concept? Or both? Front Agric Sci Eng 7:246–250. https://doi.org/10.15302/J-FASE-2020326

Renton CA, Lafave CH, Sierks K (2020) Farmers on the frontlines of the regenerative agriculture transition. Conservation Finance Network. https://www.conservationfinancenetwork.org/2020/04/15/farmers-on-the-frontlines-of-the-regenerative-agriculture-transition

Rosenbloom D, Markard J, Geels FW, Fuenfschilling L (2020) Why carbon pricing is not sufficient to mitigate climate change – and how "sustainability transition policy" can help. Proc Natl Acad Sci U S A 117:8664–8668. www.pnas.org/cgi/doi/10.1073/pnas.2004093117

Sanderman J, Hengl T, Fiske GJ (2017) Soil carbon debt of 12,000 years of human land use. Proc Natl Acad Sci U S A 114:9575–9580. https://doi.org/10.1073/pnas.1706103114

Schoeneberger MM, Bentrup G, Patel-Weynand T (eds) (2017) Agroforestry: enhancing resiliency in U.S. agricultural landscapes under changing conditions. USDA Forest Service Gen. Tech. Report WO-96

Schuur EAG, McGuire AD, Romanovsky V, Schädel C, Mack M (2018) Chapter 11: arctic and boreal carbon. In: Cavallaro N, Shrestha G, Birdsey R et al (eds) Second state of the carbon cycle report (SOCCR2): a sustained assessment report. U.S. Global Change Research Program, Washington, DC, pp 428–468. https://doi.org/10.7930/SOCCR2.2018.Ch11

Shi L, Moser S (2021) Transformative climate adaptation in the United States: trends and prospects. Science 372:eabc8054. https://doi.org/10.1126/science.abc8054

Sollenberger LE, Kohmann MM, Dubeux JCB, Silveira ML (2019) Grassland management affects delivery of regulating and supporting ecosystem services. Crop Sci 59:441–459. https://doi.org/10.2135/cropsci2018.09.0594

Sperow M (2020) Updated potential soil carbon sequestration rates on U.S. agricultural land based on the 2019 IPCC guidelines. Soil Tillage Res 204:104719. https://doi.org/10.1016/j.still.2020.104719

Stevens AW (2018) Review: the economics of soil health. Food Policy 80:1–9. https://doi.org/10.1016/j.foodpol.2018.08.005

Stott DE, Moebius-Clune BN (2017) Soil health: challenges and opportunities. In: Field DJ, Morgan CLS, McBratney AB (eds) Global soil security, Progress in soil science. Springer, Cham, pp 109–120. https://doi.org/10.1007/978-3-319-43394-3_10

Swan A, Williams SA, Brown K et al (2015) COMET-Planner. Carbon and greenhouse gas evaluation for NRCS conservation practice planning. A companion report to www.comet-planner.com. http://comet-planner.nrel.colostate.edu/COMET-Planner_Report_Final.pdf

Sweet SK, Schuldt JP, Lehmann J, Bossio DA, Woolf D (2021) Perceptions of naturalness predict US public support for soil carbon storage as a climate solution. Clim Chang 166:22. https://doi.org/10.1007/s10584-021-03121-0

Taskforce on Scaling Voluntary Carbon Markets (2021) Final report. https://www.iif.com/Portals/1/Files/TSVCM_Report.pdf

Thaler EA, Larsen IJ, Yu Q (2021) The extent of soil loss across the US Corn Belt. Proc Natl Acad Sci U S A 118:e1922375118. https://doi.org/10.1073/pnas.1922375118

The Climate Pledge (2021) Signatories. Available online at https://www.theclimatepledge.com/us/en/Signatories.html

The White House (2021) Executive order on tackling the climate crisis at home and abroad. https://www.whitehouse.gov/briefing-room/presidential-actions/2021/01/27/executive-order-on-tackling-the-climate-crisis-at-home-and-abroad/

Thompson NM, Hughes MN, Nuworsu EKM et al (2021) Opportunities and challenges associated with "Carbon Farming" for U.S. row-crop producers. Purdue University, Center for Commercial Agriculture. https://ag.purdue.edu/commercialag/home/resource/2021/06/opportunities-and-challenges-associated-with-carbon-farming-for-u-s-row-crop-producers/

Toor GS, Yang YY, Das S, Dorsey S, Felton G (2021) Soil health in agricultural ecosystems: current status and future perspectives. Adv Agron 168:157–201. https://doi.org/10.1016/bs.agron.2021.02.004

United States Department of Agriculture (2021) Climate-smart agriculture and forestry strategy: 90-day progress report

United States Environmental Protection Agency (2021) Inventory of U.S. greenhouse gas emissions and sinks: 1990–2019. EPA 430-R-21-005. U.S. Environmental Protection Agency, Washington, DC. https://www.epa.gov/ghgemissions/inventory-us-greenhouse-gas-emissions-and-sinks-1990-2018

Wagner G (2021) Recalculate the social cost of carbon. Nat Clim Chang 11:293–294

Wiese L, Wollenberg E, Alcántara-Shivapatham V et al (2021) Countries' commitments to soil organic carbon in nationally determined contributions. Clim Pol. https://doi.org/10.1080/14693062.2021.1969883

Wiggins Z, Nandwani D (2020) Innovations of organic agriculture, challenges and organic certification in the United States. Sustain Agric Res 9:50–57

Wood SA, Bowman M (2021) Large-scale farmer-led experiment demonstrates positive impact of cover crops on multiple soil health indicators. Nat Food 2:97–103. https://doi.org/10.1038/s43016-021-00222-y

World Business Council for Sustainable Development (2018) The business case for investing in soil health. www.wbcsd.org

Wuebbles DJ, Fahey DW, Hibbard KA et al (2017) Executive summary. In: Wuebbles DJ, Fahey DW, Hibbard KA et al (eds) Climate science special report: fourth national climate assessment, vol I. U.S. Global Change Research Program, Washington, DC, pp 12–34. https://doi.org/10.7930/J0DJ5CTG

Yang P, Zhao Q, Cai X (2020) Machine learning based estimation of land productivity in the contiguous US using biophysical predictors. Environ Res Lett 15:074013. https://doi.org/10.1088/1748-9326/ab865f

Yu W, Weintraub SR, Hall SJ (2021) Climatic and geochemical controls on soil carbon at the continental scale: interactions and thresholds. Glob Biogeochem Cycles 35:e2020GB006781. https://doi.org/10.1029/2020GB006781

Zomer RJ, Bossio DA, Sommer R, Verchot LV (2017) Global sequestration potential of increased organic carbon in cropland soils. Sci Rep 7:15554. https://doi.org/10.1038/s41598-017-15794-8